The Carbon Dioxide Problem

The Carbon Dioxide Problem
Integrated Energy and Environmental Policies for the 21st Century

Toshinori Kojima

Department of Industrial Chemistry
Seikei University
Japan

Translated from the Japanese Edition and Edited by

Brian Harrison

Faculty of Policy Studies
Chuo University
Japan

Gordon and Breach Science Publishers

Australia • Canada • China • France • Germany • India
Japan • Luxembourg • Malaysia • The Netherlands • Russia
Singapore • Switzerland

Originally published in Japanese in 1994 as NISANKA TANSO MONDAI USO TO HONTO by AGNE SHOFU Publishing Inc., Tokyo
© 1994 by AGNE SHOFU Publishing Inc., Tokyo

Amsteldijk 166
1st Floor
1079 LH Amsterdam
The Netherlands

British Library Cataloguing in Publication Data

Kojima, Toshinori
  The carbon dioxide problem : integrated energy and
  environmental policies for the 21st century
    1. Atmospheric carbon dioxide   2. Greenhouse effect,
  Atmospheric
  I. Title   II. Harrison, Brian
  363.7'3874

ISBN: 90-5699-127-2

Contents

Preface

The topics of environmental pollution and global warming have received much publicity in recent years. Once a substance has been identified as harmful, the most usual countermeasure employed is to control the discharge or the spread of that substance in the environment. However, in the case of carbon dioxide (which is the greatest contributor to the current problem of global warming) this approach is greatly limited since carbon dioxide is inevitably produced during the combustion of fossil fuels, and tight restrictions on fossil fuel use would have major repercussions on the economies and living standards of the industrial world. A further unique point about carbon dioxide is that unlike a number of pollutants, the substance is itself not directly harmful to living creatures or plants in the concentrations present in the atmosphere.

Although a slow warming of the earth in itself might not pose a problem (such warming has occurred in the past), the current rate of increase in global concentrations of carbon dioxide is unprecedented. Unless action is taken to remedy the situation, this could have catastrophic long-term effects.

The 'carbon dioxide problem' thus refers to the extremely complex problem of how to limit long-term carbon dioxide concentrations to levels that pose no environmental risk without devastating national economies and greatly reducing living standards. This book will argue that the key to resolving the problem lies not in treating global warming as an environmental problem but rather one of developing a comprehensive energy policy that emphasizes the need to use energy efficiently and to develop alternative sources of renewable energy. It will examine a range of technological countermeasures for the control of carbon dioxide and evaluate their feasibility and effectiveness.

In Japan the rapid post-war growth led to major pollution problems. Then in the 1970s, as a result of her great dependence on foreign oil supplies, Japan was hit especially hard by the oil shocks. These events led to Japan being at the forefront of energy conservation and anti-pollution technology, and consequently to her being in a position to make a contribution on a global scale. This book therefore provides some details of the Japanese experience in the hope that this perspective will both be of use in a general discussion of such topics and also be of interest to Western readers who otherwise might find it difficult to acquire information about Japanese viewpoints.

The Carbon Dioxide Problem attempts to fill a perceived gap in the literature by covering a very broad range of subject matter in a manner which hopefully will prove extremely useful to those looking for a general treatment of this most serious issue. It is hoped that this book will make readers more aware of the scale and nature of the carbon dioxide problem, and will thus act as a spur to a much-needed debate of an issue that is of such vital importance to the future well-being of our world.

Acknowledgements

The author would like to express his sincere gratitude to Professor Amit Chakma of the University of Regina, Canada, for carefully reading the manuscript and providing numerous valuable comments.

The author is also extremely grateful to the Japanese Ministry of Education, Science and Culture, whose financial assistance in the form of a Grant-in-Aid for Scientific Research made this project possible.

1 GLOBAL WARMING AND ITS EFFECTS

Global warming and the environmental problem are nowadays topics of great concern throughout the world. Awareness of pollution problems particularly developed during the period of post-war growth, and nowhere was this more true than in Japan. Indeed, the name of Minamata, the site of serious mercury contamination which had devastating consequences on the victims, is a household name throughout much of the developed world. Since that initial tragic encounter with the perils of ignoring the effects of human beings' actions on the environment, through to the trials of the oil shocks in the 1970s and the perceived threat to the nation's energy supplies, Japan has been at the forefront of both the fight against pollution and the development of technology to ensure stable energy supplies. Since the carbon dioxide problem spans both the areas of pollution and energy policies, it is perhaps appropriate to introduce this topic by looking at Japan's experiences. This chapter will then continue by defining the global warming problem and briefly introducing a proposal for directly cooling the earth.

1.1 JAPAN IN THE AGE OF POLLUTION

The rapid growth in the Japanese economy that followed post-war construction was accompanied by serious pollution problems. Perhaps the most infamous incidents were the appearance of Minamata disease (due to mercury poisoning) in 1953, *itai itai* disease (a painful bone disease caused by cadmium poisoning) in 1955, and Yokkaichi asthma (in Yokkaichi, Mie prefecture, due mainly to the emissions of sulfur oxides; a rapid increase in patients occurred in 1959 and the early 1960s); in 1964 the second outbreak of Minamata disease occurred in Niigata prefecture. The major sources of pollution were traced to chemical plants operated by firms such as Chisso Corporation (responsible for Minamata) and also petrochemical complexes.

In order to combat pollution, various regulations were enacted (such as the basic environmental policy law that was promulgated in 1967), and a range of procedures were implemented to ensure adherence to the regulations. Naturally, these measures included regulations concerning the recovery and removal of pollutants. In some cases, as happened in the soda industry, the process itself was changed (from the mercury method to the diaphragm method, and then to the ion exchange method). The special feature of these processes was that the

1

final product or material remained the same as before, but additional energy was used and the process was improved so as to avoid the production of pollutants. Consequently, since it was the process itself that was changed and not the end-product, no great inconvenience was caused to the consumer—except perhaps for a possible slight increase in the cost of the product. The predicament thus turned out to be an opportunity, even leading to the development of more energy-efficient processes.

However, even though the uproar which accompanied the early major pollution incidents has died away, serious problems still exist with matters such as photochemical smog and red tide (for an explanation of red tide, see the glossary). Fresh problems have also arisen in various parts of Japan, such as the contamination of the water table due to the use of agricultural chemicals at golf courses, and the recent problem in which harmful materials leaked from the waste disposal site at Hinode in Tokyo. However, it is clear that many of these cases do not stem from pollution originating directly from the manufacturing processes themselves. Instead, the main source of pollution has either been ordinary people or sites closely associated with everyday life. Problems such as vehicle exhaust fumes, sewerage and waste disposal etc. can still be alleviated by performing actions which are benign to the environment and by using goods that do not harm the earth. Each person should be aware of the consequences of their actions, and should avoid using items that are a potential cause of pollution; in addition, they need to pick up the financial burden of measures needed to treat the goods after use.

One example of pollution that is not restricted to manufacturing plants is the problem concerning nitrogen oxides,* which are a source of air pollution; these pollutants are also emitted from sites at which electricity is generated for household consumption etc. (for example, thermal power plants). In Japan, however, there has been a great reduction in the quantity of nitrogen oxides emitted, and in this respect the standards in the country are now at the highest level in the world (unfortunately, Japanese charges for electricity are also at the highest world level!). At the present time, it is estimated that the greatest challenge concerns emissions from transportation that is closely linked with everyday life, such as diesel trucks and buses.

The chief culprit with respect to waste products and pollution is ceasing to be only business logic (and possibly business ethics). If regulations become strict and difficulties are encountered with production, this will clearly pose problems not only for the enterprises concerned but will also have an effect on the ordinary person.

---

*Nitrogen oxides, often abbreviated to NO_x, is a general term that includes both nitric oxide (chemical formula NO) and nitrogen dioxide (NO_2). The term generally does not include nitrous oxide N_2O, which is one of the substances (in addition to freons) that is widely believed to be destroying the ozone layer and causing global warming (although recently a theory has been put forward suggesting that the destruction of the ozone layer by freons may actually be blocked by nitrous oxide). Along with sulfur oxides (generally referred to as SO_x), nitrogen oxides are a cause of acid rain.

This is not pollution whereby a certain small number of companies cause harm to many unspecified people; human beings destroy the human environment, which then produces a threat to people's daily lives. It is not only the sufferers who remain unspecified; and the same applies to the emitters of pollution. In 1972 the United Nations Conference on the Human Environment established the United Nations Environment Programme; with this, attention turned away from pollution itself to the environment in general, thus perhaps marking the start of a new era. However, in essence the specific programmes discussed at the conference still addressed regional (as opposed to global) problems.

Another landmark event also took place in 1972: the publication of *The Limits of Growth* by the Club of Rome. This book was the one which most accurately encapsulated the trend away from a regional perspective to a global viewpoint. The chain of events concerning environmental pollution that took place from this year on is outlined more fully in Section 5 of this chapter; first, however, let us examine the arguments presented in the book.

1.2   THE START OF THE DEBATE ON THE EARTH'S LIMITS

The publication of *The Limits of Growth* by the Club of Rome in 1972 had a major impact on the generation who were reaching adulthood at the time. The book pointed out that the earth's resources (and energy resources in particular) were not infinite, and painted the scenario of human destruction in convincing fashion. For example, consider Fig. 1.1, which is based on the assumption that the trends observable at that time in food requirements, industrial production and population would continue to rise in a geometric progression. The book warned that under such conditions the world's resources (which had been estimated as being sufficient for another 250 years at 1970 consumption rates) would in fact become rapidly exhausted. It predicted that the per capita industrial production and the output of services (health, education, banking, insurance etc.) would reach a peak in the year 2010 and would then begin a rapid decline. The book forecast that this would be accompanied by a rapid rise in mortality rates (which would return to 1900 levels by the year 2050) and decreasing populations.

Figure 1.2 shows what would happen if there were double the amount of resources. If measures taken to combat pollution proved successful, the rapid effects suggested in the earlier scenario would not occur, but population levels would still register a peak due to limits on agricultural production. That is to say, if we leave the situation to natural forces the tardiness in establishing counter-measures will still lead to the tragedy of higher mortality rates and subsequent sudden decreases in population.

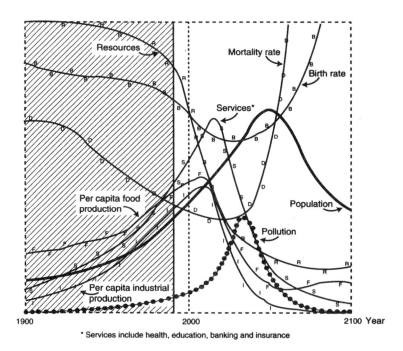

FIGURE 1.1 Model using baseline calculations [Meadows *et al.*, 1972].

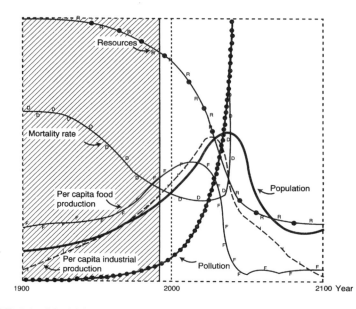

FIGURE 1.2 Model based on doubling of resources [Meadows *et al.*, 1972].

This assessment led to the concept of stabilization. However, as shown in Fig. 1.3, even if population, per capita food provisions, industrial production and services all became stabilized, the shortage of resources would finally mean that one of these parameters would no longer remain stable. Of these various situations, the most unrealistic would be the immediate stabilization of population.

These scenarios are not just hypothetical projections based on estimations of resources, various interrelationships and feedback structures* etc., but incorporate the effects of future human behavioral patterns and advances in technology. The extension of the life span of resources is technologically feasible, and would mean that products with the same function would have to be produced using a lesser amount of resources; likewise, it would be technologically possible to mitigate pollution. However, there is a limit to what can be done. For example, let us take the case in which a certain industrial production process must be accomplished using only 25% of the required resources and resulting in only 25% of the pollution. In such a case, each of the factors involved would be stabilized in the future at the desired levels without directly constraining population and by maintaining the average of two children per family (Fig. 1.4). However, the fact

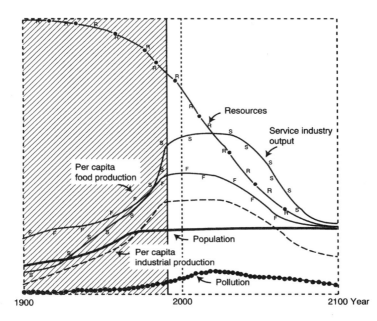

FIGURE 1.3 Model based on stabilization of population [Meadows *et al.*, 1972].

---

*If a certain cause *A* produces an effect *B*, then in turn that effect *B* will influence the cause, *A*. If the result is such that the influence of *A* is further increased, this is referred to as a positive feedback. If this effect is too great, the interaction will lead to a rapid change.

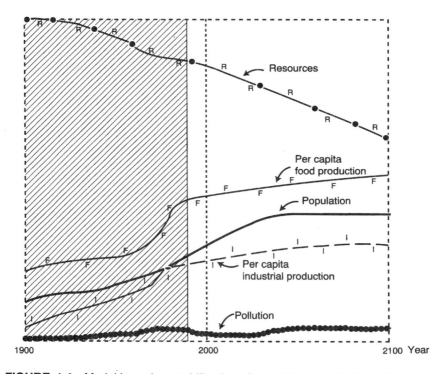

FIGURE 1.4 Model based on stabilization of population and highly efficient use of resources [Meadows *et al.*, 1972].

remains that there is one factor that simply will not be stabilized. Every single one of the hypotheses postulates that resources will become exhausted within at most a few hundred years.

The wastage of resources is a double-edged sword for the human race. Even if the environment can recover, the devouring of resources would still leave human beings on the road to ruin.

1.3 GENERAL RESOURCES MAY REMAIN UNEXHAUSTED; ENERGY RESOURCES ARE FINITE

It goes without saying that one day resources will be exhausted. However, with the exception of the consequences of nuclear reactions, the quantity of chemical elements is constant—they certainly will not disappear. If these elements are used, it simply means that they will become dispersed. The author and co-workers have termed these "diffusive" resources (Kojima and Saito, 1994). However, after dispersion, recovery becomes difficult. Indeed, in certain cases it is easier to extract materials from waste products than it is to acquire them from other resources (and this especially will be the case in the future). This means that recycling will greatly increase in importance in the future. However, if the procurement of all materials becomes difficult, this task will necessitate the expenditure of considerable amounts of manpower or energy. Fortunately, this is where technological advances are expected to provide a contribution, although at the present time, the actual amount of energy required for separation is many times the theoretical amount.

The only situation that is different is the one concerning energy resources. Strictly speaking, energy itself is being conserved. It is merely being transformed and dispersed; entropy, however, is increasing.*

Strictly speaking, the extraction of mineral resources produces a decrease in entropy; by way of compensation, the consumption of energy resources leads to an increase in entropy. As a result, it is perhaps better to talk of low-entropy resources rather than energy resources. However, leaving aside the matter of terminology, once these resources have been used, the form of their existence is transformed, and they cannot again be used as resources. The combustion of oil and coal releases energy and produces water and carbon dioxide (CO_2); it simply does not matter how much water and carbon dioxide is collected after the energy has been extracted—from the standpoint of the energy, this is irrelevant.

The increase in entropy that results from dispersion of the materials is generally small compared to the increase in entropy when energy is consumed, i.e. various forms of energy are converted into heat, thus making it possible to use energy to recover mineral resources. However, if carbon dioxide were to be used as a raw material, the conversion back into the original fuel or carbonaceous material would require the input of the same amount of energy as that which was initially liberated during the combustion process that released the carbon dioxide. If we now substitute the term entropy, it is necessary to reduce the quantity of entropy by an amount greater than the increase produced by the combustion of the fossil fuel. However, if we were to do that, the overall result would be a pointless increase in entropy.

---

*According to the first law of thermodynamics, the total amount of energy is constant. According to the second law of thermodynamics, there is an increase in the total amount of entropy (a measure of disorder) when natural phenomena occur. If fossil fuels are ignited, they will naturally undergo combustion and release energy in the form of heat; this means that for the same quantity of thermal energy produced, thermal energy alone involves greater disorder than occurs with chemical energy originating within the fossil fuel, and the value of the energy is smaller. Once converted into heat, the energy cannot be converted back into the same quantity of chemical energy; also, as the temperature approaches room temperature, there is an increase in disorder (entropy) and the proportion that can be converted back decreases.

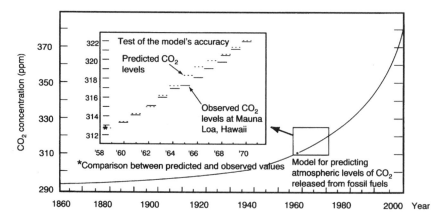

FIGURE 1.5 Increase in carbon dioxide concentrations predicted in "The Limits of Growth" [Meadows *et al.*, 1972].

Human activity results in increased entropy and depleted resources, with fossil fuels being converted into carbon dioxide. Energy resources would continue to decrease, and the concentrations of carbon dioxide would continue to rise. In fact, this is the second law of thermodynamics in operation.

Until very recently, nobody predicted that carbon dioxide would pose a problem of today's magnitude. Although the rise in carbon dioxide concentrations was correctly singled out by the Club of Rome (see Fig. 1.5), the report did not specify where the problem lay. The "fear" indicated in the report is now becoming recognized as being "reality."

1.4 THE OIL CRISES

The report by the Club of Rome was sensational, and when it was published few people thought that it described the actual situation. At least, very few felt any degree of urgency and danger at the time, even though they were fully aware that such a situation might develop some day. However, the scenario seemed to have become reality when the fourth Middle East war resulted in the oil crisis in 1973, and when that was then followed by the second oil crisis in 1979.

However, although countries such as Japan which were over-dependent on oil referred to the situation as an energy crisis, from a global perspective, it was simply an oil crisis. Certainly, the situation was not caused by a shortage of resources. Figure 1.6 shows the changes with time in the amounts of oil production and

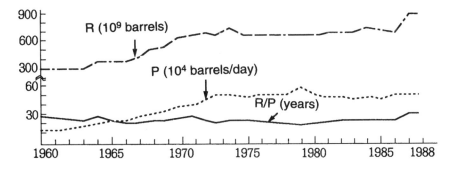

FIGURE 1.6 Changes with time in world oil production (P), proven recoverable reserves (R), and the ratio between production and reserves [Matsui, 1991].

proven recoverable reserves, as well as the ratio between them. For at least the last few decades, it has seemed natural for the human race to believe that new oil fields will be discovered, and that more than approximately 30 years of oil reserves will be at hand. But, given that there is a finite limit, how extensive are the world's ultimate reserves? The world's energy reserves are listed in Table 1.1. The ultimate reserves include those that have already been used, but it is possible that the estimates may rise substantially since they contain errors, and because new resources may be discovered which are completely different in origin to those previously known (for example, this may happen in the case of natural gas trapped deep underground).

How long will fossil fuel resources last? And will the scale of consumption continue to increase? Even though various factors are involved, for simplicity let us consider the values produced by dividing the estimated ultimate reserves by the amount of production. According to Table 1.1, the total reserves of oil and natural gas should both last for 100 years. Of the fossil fuel consumption, oil accounts for 43% and natural gas for 23%. Thus the two of them together account for two-thirds of all fossil fuel use, with the other one-third accounted for by coal. If oil and coal were to be replaced by natural gas, supplies would last for 20 years. On the other hand, supplies of coal would last for over 2,000 years at current levels of production. It was thought that if people exhausted the supplies of oil and natural gas, it would be acceptable to replace them by the use of coal. In the end this would mean that total fossil fuel supplies would last for 700 years; the figure would be over 800 years if oil sand and oil shale were also included. As a result, it was believed that even if energy resources are considered to be finite, the situation would be satisfactory if new forms of renewable energy could be developed within a few hundred years (this will be examined in greater detail in Chapter 3). However, the emergence of the global warming problem means that this conclusion was probably over-optimistic.

Table 1.1 World energy reserves (Matsui 1991).

	Type of Resource				
	Oil (10^10 barrels)	Natural gas (10^12 m^3)	Coal (10^9 t)	Oil sand and oil shale (10^10 barrels)	Uranium (10^4 t)
Ultimate reserves	200	204	9,900 (High-grade coal=6,900)	Oil sand 160 Oil shale 550	Unknown
Proven recoverable reserves (R)	90.74 (end of 1988)	112 (end of 1988)	730.8 (High grade coal=515.8) (October 1986)		230 (Under $130/kgU) (January 1988)
Existing reserves by region [%]			High-grade coal		
North America	3.7	7.1	26.3	74.0	26.4
Central and South America	13.4	6.0	0.5	21.1	8.3
Western Europe	2.0	5.1	6.2		8.0
Middle East	63.0	29.9	0.2		0.2
Asia and the Pacific	2.4	6.1	7.9	other regions=4.9	25.5
Africa	6.3	6.4	12.4		31.6
Former communist bloc	9.2	39.5	46.5		Unknown
Annual production (P)	2.11 (1988)	2.01 (1988)	High-grade coal 3.18 (1986)	Small amount	3.7* (1986)
Time until reserves are exhausted (R/P)	43 years (All world)	56 years (All world)	High-grade coal 162 years (All world)	Many years	40.63 years*

*not including former communist bloc.

1.5 GLOBALIZATION OF THE POLLUTION PROBLEM

Various effects of the global environmental problem have become apparent, such as changes in aquatic environments and climate etc. Specifically, the repercussions have been evident in desertification, marine pollution, nuclear waste contamination, acid rain, destruction of the ozone layer, and the transfer of harmful substances over international borders; the environmental problem has been recognized as one that encompasses population levels, food production, energy and resources. With the minor exception of a few freak accidents of nature, all of these harmful effects have stemmed from the activities of human beings. It is clear that the origins of the problem lie in the development and consumption plans which were devised by the advanced nations even though they harmed the earth, and in the population problems that afflict developing countries. Despite this deep flaw, the advanced nations continue to pursue the same growth-oriented policies, and the developing countries have also exhibited the desire to follow the same "road to prosperity" as the advanced nations.

In exactly the same way as has occurred with the pollution problem so far, resolution of some of the difficulties would be possible if existing technology could be applied (or further developed), or if the production and use of certain materials were ceased and other materials used as a substitute. The question of whether this can actually be achieved depends on human wisdom and a recognition of the importance of the problem, but in theory it is certainly possible. This applies to all the various problems of ozone layer destruction (a problem related to the release of freon gases), acid rain, marine pollution, and nuclear waste contamination. Although various individual difficulties remain, such as the development of a desulfurization method appropriate for developing countries, it should be possible to resolve them by the application of both general wisdom and investment. Most of the sources of emission of the pollutants have been identified, and the technology for treating them is available. Unfortunately, since the application of this technology results in a slight drop in efficiency and a small rise in cost, the technology has not been adopted. In a sense it is a problem of organization and recognition. Certainly, in Japan's case these difficulties have been overcome. The world is hoping that this experience will be put to good account in combatting global environmental problems.

Related to the problems of overpopulation and development in the developing world are the issues of desertification, depletion of tropical forests, and food production—topics which form a great barrier in the North–South dialog (i.e. the dialog between the advanced and developing nations). This again is a problem that can be solved—indeed, *must* be solved—by a combination of understanding and recognition; this means that the advanced nations must make efforts in order to restrain development and to engage in recycling and not

Table 1.2 Significant events affecting global environmental issues.

March 1972	"The Limits to Growth" published by the Club of Rome
June 1972	United Nations Conference on the Human Environment, Stockholm
December 1972	Establishment of United Nations Environment Programme (UNEP)
August 1977	United Nations Conference on Desertification, Nairobi (Kenya)
May 1984	World Commission on Environment and Development (WCED)
March 1985	Vienna Convention for the Protection of the Ozone Layer (by United Nations Environment Programme, UNEP)
June 1985	Tropical Forestry Action Plan (Food and Agriculture Organization of the United Nations, FAO)
July 1985	Helsinki Protocol: Protocol on the reduction of sulfur emissions or their transboundary fluxes by at least 30 percent (Helsinki, Finland)
April 1986	Nuclear accident at Chernobyl, Ukraine, Soviet Union
February 1987	Final meeting of World Commission on Environment and Development (WCED)
April 1987	Publication of "Our Common Future" (WCED)
September 1987	Montreal Protocol on Substances that Deplete the Ozone Layer ("Montreal Protocol")
June 1988	Toronto Conference on "The Changing Atmosphere" produced a 'call for action' on energy and environmental questions
October 1988	Sofia Protocol: Protocol concerning the control of emissions of nitrogen oxides or their transboundary fluxes (Sofia, Bulgaria)
November 1988	Establishment of the Intergovernmental Panel on Climate Change, IPCC in Geneva, Switzerland (by UNEP and WMO, World Meteorological Organization)
March 1989	Basle Convention on the Control of Transboundary Movements of Hazardous Wastes and Their Disposal, Basle, Switzerland
November 1989	Noordwijk Declaration (Netherlands). Ministerial-level participants agreed a declaration on "atmospheric pollution and climatic change".
January–February 1991	Gulf war (crude oil spills, torching of oil fields)
June 1992	United Nations Conference on Environment and Development, Rio de Janeiro, Brazil. Commonly referred to as the 'Earth Summit'. (Rio Declaration on Environment and Development, Agenda 21; included agreements on forest preservation, biodiversity, climate change and global warming).
March 1994	Framework Convention on Climate Change came into force (following ratification of agreement at UNCED, Rio, 1992)

just disposal, and furthermore the developing countries must strive to solve the problems of education and overpopulation.

Until advanced countries eliminate waste and extravagance and pass on these savings to the developing world, and until the population growth in developing countries is curbed, it will be impossible to achieve sustainable development. However, the more important point is that energy sources are intrinsic to growth; indeed, they are necessary just to maintain the present state of affairs. In addition, as indicated in Sections 1.3 and 1.4, the majority of the energy used today is not sustainable. The times, and indeed the earth's environment, are demanding a halt to this.

There have been many significant pollution incidents and environmental conferences since the publication of *The Limits of Growth* (Table 1.2). One notable gathering was the inaugural meeting of the Intergovernmental Panel on Climate Change (IPCC), which took place in 1988. It seems to have been after this that the fear which was evident at the time of the oil crisis and the recognition that the earth's resources were finite began to spread once again among ordinary people. However, on this occasion the worry concerned not the energy resources but the earth's environment, and the problem was expected to develop not in a few centuries but rather during our own children's lifetime.

1.6 THE SUBJUGATION OF THE ENVIRONMENT TO HUMAN NEEDS—AND THE PROBLEMS STEMMING FROM THIS

True nature no longer exists in many of the world's cities; and even the forests show the mark of humans. The land's surface has been completely transformed by the presence of human beings; even fields and rice paddies have been carved out of the forests. After so much has happened why make a fuss about it? Does it really matter if the world's last remaining tropical forest is destroyed?

On the opposite side of the coin is the question of the greenification of the deserts. Leaving aside the question of whether greenification is actually possible, from the point of view of the earth this may just be another nuisance. If afforestation proceeded and the deserts disappeared, the earth would look very different, and the forms of life that live in the unique desert environment would become extinct. From the standpoint of human beings, however, this greenification seems to be extremely attractive.

The global environmental problem stems from the fact that human beings need a certain environment in order to sustain their lives, but it is the very actions of human beings themselves that is destroying this environment. This produces a range of harmful effects; for instance, people become ill and the mortality rate rises, or else the environment which people enjoy ends up being destroyed. This sequence of events is not perceived directly as a pollution problem; and in addition, the expense of either preventing the pollution or of repairing the damage is extremely high. In some cases, the situation cannot simply be rectified, such as when an entire species becomes extinct, resulting in a loss of part of the human race's heritage. The view seems to be that it does not matter how many species disappear from the face of the earth, as long as the situation does not lead to the destruction of human beings' environment. The existence of this view suggests that it would be better to treat this as separate to an environmental problem.

The loss of forests has both local and global implications. The deforestation of mountainous areas has led to a loss in the soil's ability to retain water, which in turn has probably been the cause of devastating floods that have directly produced immeasurable damage. As will be discussed later, there is a growing acknowledgment that this deforestation has also had a considerable effect on the problems associated with carbon dioxide. In this sense it could be said that there is a growing recognition that deforestation is part of a broad environmental problem that is of worldwide proportions.

Table 1.2 shows one aspect of the pollution question that has not yet been addressed—that of accidents and wars. The main example that springs to mind is the nuclear accident that occurred at Chernobyl in the former Soviet Union in 1986, which one fervently hopes will never be repeated. Another instance was the discharge of oil and the torching of oil fields during the Gulf War in 1991. Then in the early part of 1993 the oil tanker Braer ran aground on the coast of the Shetland Isles (situated to the north of the British mainland), spilling its cargo of crude oil into the ocean. Around that time, in the space of about just one month several accidents occurred involving oil tankers. We will not go into a detailed discussion of these in terms of their being part of the global environmental problem since such accidents and wars are basically preventable. Naturally, though, we must attempt to devise infallible policies that will preclude such events ever occurring again.

1.7 SPECIAL FEATURES OF THE GLOBAL WARMING PROBLEM

Let us now consider the special characteristics of the global environmental problem referred to in Section 1.5. First, let us look at materials which have either been recognized as being essentially harmful or which are waste products. Either intentionally or because of negligence, they are left untreated, dispersed or transported elsewhere. The effects are seen in worldwide contamination, various forms of marine pollution, nuclear waste contamination, and the transport of harmful materials across international borders. By acting in a more responsible

fashion, taking greater care, and implementing safety measures more conscien-
tiously, it should be possible to prevent these forms of pollution. As in the case
of wars and accidents, they are environmental problems that should never be
allowed to happen.

Next, let us consider environmental problems which used to be localized ones
but which then developed into global problems; examples include acid rain, and
some forms of marine pollution. In a sense the countermeasures that need to be
taken against these are well-defined, but in some cases are not put into practice
due to the domestic conditions within a certain country or because of economic
reasons. This is an area in which Japan should play a major role due to her status
as an advanced nation, both in terms of her record concerning pollution history
and her experience with anti-pollution policies. In contrast, the destruction of
tropical forests and desertification (which are problems related to the planting of
vegetation in the peripheral areas of tropical forests and deserts) are in many
cases rooted in the overpopulation problem, which in turn produces the problem
of inadequate food production. The advanced nations have also played a major
role in the destruction of tropical forests, but it is a problem that faces the whole
human race since people only evaluate forests in terms of their traditional economic value rather than in terms of their true non-economic value (i.e. the value
of their environmental role in flood prevention and in support of ecosystems,
and their role in the carbon dioxide problem). Even so, the causal relationship
is clear, and if human beings seriously appreciate that value, it is unlikely that
anyone would oppose countermeasures to combat the threat.

Unlike the issues discussed so far, the questions of the destruction of the
ozone layer and of global warming are characterized by the fact that the causal
relationship can only be clarified by scientists. Just as the world includes many
people who do not believe in religion, so it also includes many people who are
either sceptical about science or simply do not want to put credence in it. It
comes down to the fact that people will simply not want to believe something if
they realize that a loss will be incurred. A typical example is provided by freon
gases. Freons are chemical substances which are harmless to human beings and
which used to be regarded as extremely useful substances. Therefore, despite the
evidence, the view that freon gases were a cause of the problem was for a long
time not widely acknowledged.

Global warming results from the actions of so-called "greenhouse gases".
These gases are typical of a global environmental problem in that their sources
are widely dispersed, virtually all of them are harmless to humans, and their
causal relationship with global warming remained unclarified for a long time.
Furthermore, in the case of carbon dioxide, in addition to the large quantities
involved and the difficulty of specifying the sources of emission was the contribution of a range of other factors such as the North–South problem, and the
problems of overpopulation and food production.

1.8   CAUSES OF GLOBAL WARMING

It is still unclear whether greenhouse gases do actually cause global warming, and whether global warming would actually constitute a major problem. Here we will only consider the essence of this debate.

First it is necessary to dispel a number of misunderstandings. To start with, there is the question about whether the earth is being warmed as a result of the heat released due to energy use. The total annual amount of energy used by human beings is equivalent to about 8×10^9 tons of oil. Since 1 kg of oil is the equivalent of approximately 10,000 kilocalories (kcal), this amounts to a total of 8×10^{16} kcal (or about 33×10^{16} kilojoules, kJ). However, even if all of this were to be used to warm the oceans, the annual temperature rise would be a mere 6×10^{-5}°C. Even over a period of 10,000 years the total temperature increase would fail to reach 1°C. The energy used globally in one day is a little short of 0.1×10^{16} kJ, whereas the daily amount of solar energy received by the whole earth is $1,500 \times 10^{16}$ kJ. In other words, the amount of energy released each day by human beings is less than one ten-thousandth of the energy received from the sun. Even if the earth's surface is warmed each day during daytime, it cools rapidly when the sun goes down—a pattern that is repeated every day of the year. Even if the difference between day and night temperatures increases by one ten-thousandth, it would not appear to be particularly important, and seems much less than when there is a change of season.

But now let us examine the situation more closely. Since the earth receives thermal energy only from the sun, the heat emitted from the earth increases by one ten-thousandth.

The amount of heat that is released from a body is directly proportional to the fourth power of the absolute temperature of the body. Therefore, in order to increase the amount of heat released by one ten-thousandth, it would be necessary to raise the absolute temperature of the body (in this case, the earth) by one forty-thousandth.* Since the present temperature of the earth is just under 300 K,

---

*The relationship between energy and absolute temperature is given by:
$$E = kT^4$$
where E = energy, T = temperature (in Kelvin, K), and k is a constant of proportionality. This is known as the Stefan–Boltzmann equation.

When the initial temperature is given by T_0, and this is increased by a factor of $(1 + \alpha)$, the new temperature, T, is given by:
$$T = T_0(1 + \alpha)$$
At this time, the radiated energy is given by:
$$E = kT_0^4(1 + \alpha)^4$$
This is approximately equal to
$$E = kT_0^4(1 + 4\alpha)$$
In other words, the energy increases by $(1 + 4\alpha)$ times.

the temperature would be raised by the order of only 0.01 K. This would cover less than 1% of the effect of the greenhouse gases.

If calculations are performed using the Stefan–Boltzmann equation described in the footnote, the earth's temperature would drop to −18°C.* However, in reality the average temperature of the earth's surface is approximately +15°C, which is due to the greenhouse effect. If the greenhouse gases were not present, there would probably be no life on the earth.

When light is radiated from a high-temperature body such as the sun, it is in the form of short-wavelength radiation such as visible light and ultraviolet rays. However, after this radiation is absorbed by the earth, which is much cooler than the sun, it is radiated back into space in the form of long-wavelength radiation such as infrared rays. In greenhouses, visible light and ultraviolet rays readily pass through materials such as vinyl and glass; however, this is not the case with infrared rays, which therefore become trapped with an accumulation of heat that causes the inside of the greenhouse to become warm. Greenhouse gases such as carbon dioxide act in the same manner as the vinyl and glass in a greenhouse, thereby raising the temperature of the earth.

The question which then arises is the extent to which the temperature of the earth's surface will be raised by an increase in such greenhouse gases. There are three points to make in relation to this. First, there is no change in the situation regarding the fact that from outside the earth's outer atmosphere (i.e. in space) the earth's temperature appears to be −18°C. Second, even if the greenhouse effect increases, the temperature of the earth's surface will not continue to rise; it would become stabilized at a certain temperature which would be determined by the concentrations of the greenhouse gases in the atmosphere. Under these stabilized conditions, referred to as a steady state, all the heat that the earth's surface receives from the sun would be transported to the cold outer atmosphere, from where it would then be radiated into space. (Of course, it is important to know the temperature of the earth's surface at which this would happen.) Finally, the greenhouse gases prevent the transport of heat from the warm surface of the earth to the cold outer atmosphere. In order to overcome this increased blocking and allow the transport of the same amount of thermal energy as has been received from the sun in the past, there needs to be a greater temperature difference between the earth's surface and the outer atmosphere. In other words, the earth's surface would need to become warmer. In the model depicted in Fig. 1.7, instead of the heat accumulating at the earth's surface as a result of the greenhouse gases it would be better to consider that (since the quantities of heat entering and leaving the planet are equal) in order to avoid a net accumulation of heat, the temperature of the earth's surface would rise, leading to an increased

---

*This calculation is based on the assumption that the earth is a black body, i.e. that it is a perfect radiator of heat. Although the calculation is only an approximation, the results yielded are for the most part valid.

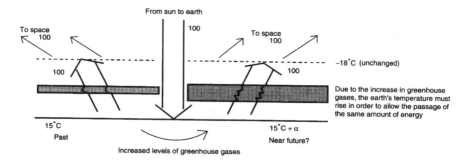

FIGURE 1.7 Principles of global warming and the greenhouse effect.

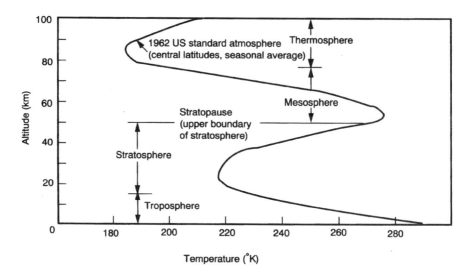

FIGURE 1.8 Vertical temperature distribution in the 1962 US standard atmosphere [*adapted from* Tanaka, 1991].

discrepancy between the actual temperature and the apparent temperature as viewed from space.

The actual temperature distribution within the earth's atmosphere is shown in Fig. 1.8 (although strictly speaking it varies slightly with latitude). Measurements show that the temperature of the stratosphere is high due to the absorption of ultraviolet rays. Looking at the temperature distribution from space, the temperature still appears as if it were $-18°C$, and this would remain unchanged no matter how great the accumulation of greenhouse gases. Thus the only change that would actually take place would be the increase in temperature at the earth's surface (or, rather, an increase in temperature gradient near the earth's surface).

In fact, according to some analyses, while the atmosphere close to the earth's surface (i.e. the troposphere) would indeed become warmer at this time, the outer parts of the atmosphere (i.e. the stratosphere) would actually become cooler).

1.9 EFFECT OF GREENHOUSE GASES ON ATMOSPHERIC TEMPERATURE

The role of greenhouse gases is actually complex and cannot be calculated in such a simple manner. Accordingly, even the Intergovernmental Panel on Climate Change (IPCC, 1990) produces a range of estimates. It has been calculated that a doubling of carbon dioxide concentrations would lead to an increase in temperature ranging anywhere from 1.5°C to 4.5°C (Fig. 1.9). The calculations would not be so difficult if the actions of only the gases conventionally regarded as greenhouse gases were taken into account. However, a rise in atmospheric temperature would lead to other complicating factors such as an increase in water vapor concentrations. Water vapor also behaves in the same manner as conventional greenhouse gases and would thus normally act to raise the temperature of the earth's surface, but as water vapor produces clouds (which form a barrier to solar radiation) it also acts in such a way as to lower the temperature of the earth's surface. It is unclear as to which of these two opposing actions has the greater effect.

Sulfur compounds also act as greenhouse gases; but since these easily lead to the formation of aerosols, which block the sun's rays, they could also have a cooling effect. (Of course, these sulfur compounds also cause environmental damage since they are a source of acid rain.)

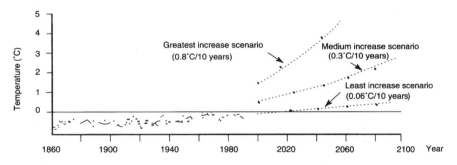

FIGURE 1.9 Changes in average temperature at the earth's surface for various scenarios [Jäger, quoted in OECD 1992] (A similar prediction was also published by IPCC, 1990).

The greenhouse effect depends on the absorption of certain specific wavelengths of light. Accordingly, if complete absorption occurs at those wavelengths, the greenhouse effect would no longer take place. The greenhouse effect per unit concentration of carbon dioxide is small compared with that of other trace gases, which can be deduced from the fact that the present concentration of carbon dioxide in the atmosphere is already more than 300 parts per million (ppm*), and since considerable absorption already occurs at its characteristic wavelength. Of course, the absorption band of carbon dioxide overlaps with that of other gases. The absorption spectrum for long-wavelength gases emitted from the earth (Fig. 1.10) shows that part of the radiation which should have been emitted from a body at 300 K has already been absorbed by greenhouse gases.

In addition, the length of time for which each gas remains in the atmosphere varies. Consequently the degree of the observed effect varies with the period selected for evaluation. The situation is often referred to as a calculation of carbon dioxide concentrations, but even this is not simple to determine.

There is one view that holds that warming has occurred because the concentrations of carbon dioxide and other greenhouse gases has risen. Figure 1.11 shows the changes in carbon dioxide concentrations from 160,000 years ago up until the present day. As indicated, on a time scale of several tens of thousands of years there is a clear and extremely good correlation between carbon dioxide

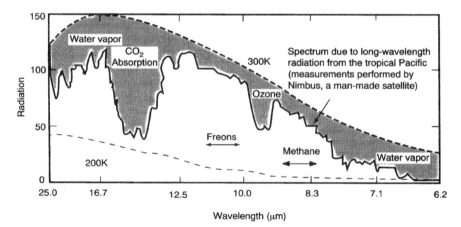

FIGURE 1.10 Spectra for long-wavelength radiation from the earth (*solid line*), and for radiation from a black body at both 300 K (27°C) and 200 K (−73°C) (*broken lines*) [Ramanathan, 1987].

*Parts per million (ppm) is a unit that describes concentration. In the case of a gas, the concentration is indicated in terms of either the volume or the number of moles. (One mole is approximately 6×10^{23} molecules, and occupies 22.4 liters at 0°C and atmospheric pressure.)

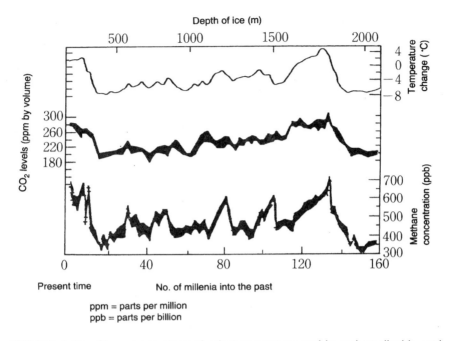

FIGURE 1.11 Changes in atmospheric temperature and in carbon dioxide and methane concentrations in the Antarctic over the past 160,000 years [Barnola *et al.*, 1987].

concentrations and temperature. However, we must not be deceived by the argument that temperature will necessarily rise with an increase in carbon dioxide.

Methane also exhibits a good correlation with atmospheric temperature, so let us now examine whether methane also is a cause of warming, and if so, why the levels of methane and carbon dioxide are interrelated. Of the three factors involved (i.e. methane, carbon dioxide and temperature), which is the cause and which is the effect? This is a little like asking which came first, the chicken or the egg.

Let us consider Fig. 1.11. With the exception of the last hundred years, carbon dioxide concentrations rose from 200 to 290 ppm over a period of 20,000 years. During that time, atmospheric temperature rose by nearly 10°C; a change of the same magnitude occurred between 140,000 and 160,000 years ago.

On the other hand, although the carbon dioxide concentration rose from 290 to 350 ppm over the past 100 years, the temperature changed by less than 1°C. Naturally most of this rise in carbon dioxide concentrations is clearly due to the combustion of fossil fuels (a point which will be further addressed in Chapter 2). Therefore, no matter what the reason might be, it seems convincing to explain the data in Fig. 1.11 as temperature being the cause and the carbon dioxide and methane concentrations being the effect. (In actual fact, the solubilities in sea water of both carbon dioxide and methane decrease as the temperature rises, with a resulting increase in their atmospheric concentrations.) The main cause of the temperature change is believed to be solar activity, which is manifested in the number of sunspots (Fig. 1.12). Over a slightly longer term, the rise in temperature is likely to be affected by factors such as modifications of the earth's orbit about the sun.

Nevertheless, carbon dioxide does indeed have an effect on atmospheric temperature. It was just that in the past the effect of the sun was greater than the effect of carbon dioxide. Everybody knows that the effect of the sun is still great; after all, it is cold in winter even though the carbon dioxide levels increase. Therefore let us now examine the effect on the long-term average temperature.

Certainly over the long term, the earth's temperature has exhibited fluctuations of $\pm 10°C$ due to the effects of the sun. However, this has occurred over a period of 20,000 years. We have already witnessed a change in carbon dioxide concentrations equivalent to that in just the last one hundred years, and it is expected that the concentrations will continue to increase. The doubling of carbon dioxide concentrations that was incorporated into the previously mentioned IPCC model is an abnormal degree of change. Under these circumstances, it is necessary to consider that the temperature of the atmosphere will rise by 1.5–4.5°C. The problem is not that an increase will take place on such a large

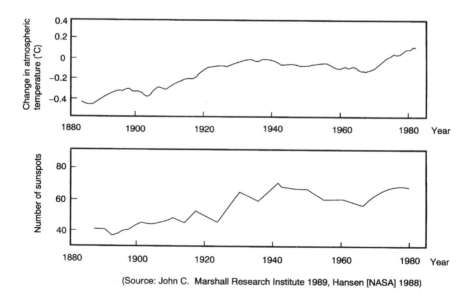

(Source: John C. Marshall Research Institute 1989, Hansen [NASA] 1988)

FIGURE 1.12 Temperature changes in the past 100 years (taking the 1951–1980 average temperature as the baseline), and changes in the number of sunspots [Ishikawa, 1990].

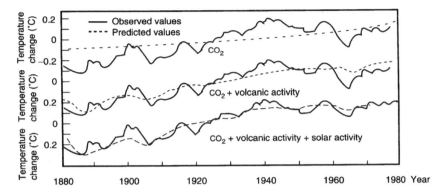

FIGURE 1.13 Predicted and observed changes in the earth's temperature (assuming a temperature rise of 2.8°C for a doubling of CO_2 concentrations) [Hansen *et al.*, 1981].

scale, but rather that it will occur in a matter of just one to two hundred years, and that such a sudden change in the earth's environment will have a profound effect on human activities.

Furthermore, some people have expressed worries that the rise in temperature will have a positive feedback effect on the carbon dioxide levels. However, based on the above argument, and assuming that past results are still valid, the effect would be of the order of 10 ppm/1°C; if so (and admittedly this is perhaps taking a rather optimistic view), the effect does not appear to be so large.

Various people have constructed a number of theoretical models to forecast the "recent" temperature changes (or perhaps a more accurate way of expressing this would be to say that they are enough to explain the past changes). One such model is shown in Fig. 1.13, which shows the long-term effect of the greenhouse gas concentrations (mainly carbon dioxide), and the effect of the sun at 10-year intervals. It has also to be remembered that volcanic activity has a cooling effect only in the short term due to aerosols and dust particles which block the sun's rays. The best fit with observed readings seems to be provided by theoretical values derived from these three parameters (i.e. carbon dioxide, volcanic activity and the sun).

1.10 EFFECT OF TEMPERATURE CHANGES ON THE EARTH'S ENVIRONMENT AND POSSIBLE COUNTERMEASURES

Let us now consider the question of why it is unacceptable for temperatures to rise. Is it really a bad thing? Overuse of air-conditioning systems will lead to a positive feedback effect on global warming. This might be welcomed by people living in cold regions, but vegetation re-germinates only in places suited to that particular form of plant life and is unable to cope with drastic changes. The same would apply to agricultural produce, although to a certain extent this could be alleviated due to the experience that humans have acquired with respect to farming. It may even be possible that biotechnology techniques such as the use of heat-resistant genes will be developed to combat these effects. It is possible, though, that certain products that grow in cold regions would become more expensive.

One manifestation of the concerns related to a positive feedback effect is that the large amount of methane which is trapped in the icefields of Siberia would be released when the ice melts. All the following examples are other possible events that may take place if temperatures rise, with the uncertainty involved being greater than the probability of global warming accompanying a rise in carbon dioxide levels.

First, let us consider the question of a rise in sea levels. It was said that the Antarctic ice would soon cause sea levels to rise by several meters by calving from the icepack and flowing into the ocean. Recently, observations in the Antarctic, where the ice has piled up on the continent of Antarctica, has led to the conclusion that these effects will not occur so rapidly. In contrast, even if the Arctic ice melts, there would be no rise in sea levels. This is analogous to the situation in which, if ice is present in a glass of whisky that is filled to the brim, water does not spill over the sides even when the ice melts. It has been postulated that the first effect that would be seen if sea temperature rises would be an increase in volume. However, it was predicted that the rise in sea level during these several decades would be not several meters but only about one-tenth of that. Notwithstanding this, the effect has been clearly visible to the eye.

The oceans have a substantial thermal capacity. Thus there would be a considerable lapse of time between a rise in atmospheric temperature and a corresponding increase in sea temperature, and in addition, any sudden change in atmospheric temperature would to some degree be suppressed due to the presence of the oceans. Nevertheless, it is indisputable that sea levels have risen by approximately 10 cm in the last hundred years (Fig. 1.14). As evidenced by events in the Netherlands and elsewhere, this rise in sea level is the largest observed in history; however, the real problem is the suddenness and acceleration of the change, rather than the absolute amount. In particular, the impact on a country which already has a major problem with sea levels would be especially great, such as in the Netherlands. Certainly the financial cost of one of the possible countermeasures (the construction of necessary dykes) would be considerable; another possible strategy, relocation to an inland area where sea water could not invade, would entail a prohibitive economic loss. The various areas around the world which would have great concern about a rise in sea levels are shown in Fig. 1.15.

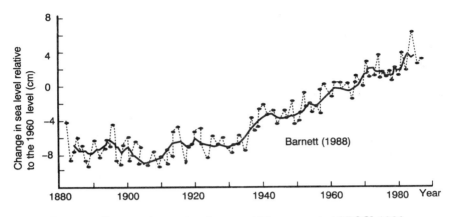

FIGURE 1.14   Changes in sea level over a 100-year period [IPCC] 1990.

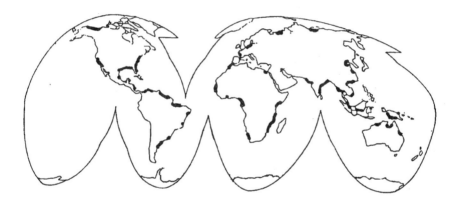

FIGURE 1.15   Areas which would be affected by a rise in sea levels [*adapted from* Rosenberg *et al.*, 1989].

Besides simply the rise in sea levels, the salt content of the water would pose additional difficulties. In the Netherlands the influx of salt water into estuaries has already presented a problem, and this scenario is likely to repeat itself in other areas. The salt water enters the underground water supply, which in turn increases the salinity of the soil; this then has a major effect on agriculture e.g. for the supply of water for irrigated land.

A further problem posed by a rise in sea level is that encroachment occurs in areas with abundant ecosystems such as deltas. If mangroves, which grow at the edge of waterways, are destroyed, the amount of carbon that is fixed in them will decrease, carbon dioxide will be released, and global warming will be accelerated still further. However, such areas are by their very nature fertile land, and some such areas have been used by people for agriculture and/or the establishment of industrial zones. On the other hand, as far as the oceans are concerned, places such as sand bars and coral reefs would sink deep into the sea and be destroyed (coral is only able to exist in the surface layers of the sea).

The effects of global warming are also clearly evident in climatic changes, due to their aggravation of various natural phenomena. A rise in atmospheric temperature increases water vaporization, resulting in more frequent and more ferocious typhoons. While this is perhaps stating the obvious, feasible countermeasures would be to construct sturdy housing, and to remain indoors when a typhoon hits.

An increase in precipitation might at first appear to be beneficial from the point of view of food production, but if (as mentioned above) there should be an increase in torrential downpours, this would create other worries such as land being washed away. The reverse scenario would be that a rise in atmospheric temperature would also increase the dispersion of water vapor, which is a cause of land becoming parched. It is actually unclear as to which areas would experience greater precipitation and which would suffer desertification. However, it would be a tragedy if an area that had just been converted into agricultural land was later turned back into desert. One countermeasure to alleviate the situation could be the development of a system to predict changes in rainfall distribution. Biotechnology could also play a role in developing vegetation that is resistant to arid conditions. In addition, rain could be induced using artificial means. However, it must be admitted that these measures seem to be only a remote possibility. We therefore need to consider alternative measures that are both positive and realistic.

1.11 COMBATTING GLOBAL WARMING BY COOLING THE EARTH

When devising a strategy to achieve a stated objective, it is important to consider all potential outcomes of the chosen course of action. Thus when considering the cooling of the earth, it would be necessary to make absolutely sure that this would indeed lead to mitigation of the adverse effects of global warming.

Some people claim that the earth could be cooled by utilizing the cold water from deep oceans. Let us consider the implications of this. It would certainly suppress global warming in the short term. However, as previously mentioned, instead of heat accumulating at the earth's surface due to the action of the greenhouse gases, the situation should perhaps be thought of as the warming of the earth's surface in order to avoid the accumulation of heat; from this perspective, the use of cold water for the purpose of cooling is in the long run nothing more than the overall accumulation of heat by the earth. If the countermeasure were discontinued, everything would return to the original situation. In other words, the earth would become warmer, and thus the technique would essentially fail in its purpose of preventing global warming. Furthermore, one effect of raising cold deep water to cool the earth would be that the temperature of this deep sea water would rise, causing expansion; this might further accelerate the rise in sea level (which itself is one of the adverse effects of global warming).

Another method of cooling the earth would be to raise the albedo.* If this should occur, energy could be released outside the earth's atmosphere without being obstructed by greenhouse gases. The areas on land with large albedos are icefields and deserts. Thus an increase in barren land could act as a countermeasure against global warming.

The reason for the cooling action of volcanic activity on the earth is that materials released from volcanoes block the sun's rays. It is believed that several sulfur compounds produce a similar effect, including sulfur dioxide. Thus the release into the atmosphere of substances that cause acid rain could actually constitute a countermeasure against global warming. However, it is unlikely that there would be much support for putting such a scheme into practice since it is quite conceivable that such gases could themselves at any time turn into being causes of environmental pollution, such as acid rain.

In conclusion, at the present time there simply does not appear to be any strategy which can be strongly recommended.

References

Barnola, J. M., D. Raynaud, Y. S. Korotkevitch and C. Lorius. 1987. *Nature* **329**, 408. Cited in Akimoto, H. 1992. *Petrotech*, **15**, 504 (in Japanese)

Hansen, J., D. Johnson, A. Lacis, S. Lebedeff, P. Lee, D. Rind and G. Russell. 1981. *Science* **213**, 957. Cited in Komiyama, H. *et al.* (eds.). 1990. *A Handbook of the Global Warming Issue*, p. 511, IPC, Tokyo, Japan (in Japanese)

Intergovernmental Panel on Climate Change (IPCC). 1990. Houghton, J. T., G. J. Jenkins and J. J. Ephraums (eds.). *Climate Change, The IPCC Scientific Assessment*. p. 364, Cambridge University Press, Cambridge (cited in Tanaka, M. 1991. In Organizing Committee of 5th Symposium on University and Science (ed.). *Science of Global Environmental Change*. p. 167, Kubapuro Pub., Tokyo, Japan

Ishikawa, K. 1990. *Petrotech*, **13**, 728

Kojima, T. and K. Saito. 1994. *Recovery and Recycling of Resources*. In Garside, J. and S. Furusaki (eds.). The Expanding World of Chemical Engineering. Gordon and Breach Science Pub., Yverdon, Switzerland

Matsui, K. 1991. *How to Read and Use Energy Data*. Denryoku Shinpo Pub., Tokyo, Japan (in Japanese)

*Here, albedo refers to the proportion of incoming light or radiation that is reflected by the earth's surface.

Meadows, D. H., D. L. Meadows, J. Randers and W. W. Behrens III. 1972. *The Limits to Growth. A Report for The Club of Rome's Project on the Predicament of Mankind.* Universe Books, New York. (Citations from the Japanese edition, 1972, Diamond Pub., Tokyo, Japan)

Organization for Economic Cooperation and Development (OECD). 1992. *The State of the Environment.* (Citations from the Japanese edition, 1992, p. 17, Chuo Hoki Pub., Tokyo, Japan)

Ramanathan, V. 1987. *Journal of Geophysical Research* **92**, 4075. Cited in Komiyama, H. *et al.*, 1990. *A Handbook of the Global Warming Issue* p. 30, IPC Publishers, Tokyo, Japan (in Japanese)

Rosenberg, N. J., W. E., Esterling III, P. R. Coroson and J. Darmstadier (eds.). 1989. *Proceedings of Workshop on Climate Change.* Washington, D.C., June 14–15, 1988. Cited in Kaya, Y. 1991. A Handbook of Global Environmental Engineering, p. 547, Ohm Pub. Inc., Tokyo, Japan (in Japanese)

Tanaka, H. 1991. In Kaya, Y. (ed.). A Handbook of the Global Environmental Engineering (in Japanese), p. 70, Ohm Pub. Inc., Tokyo, adapted from U.S. Standard Atmosphere, 1962 by NOAA, NASA and USAF, Washington D.C.

2 REDUCTIONS IN GREENHOUSE GAS EMISSIONS, AND THE EARTH'S CARBON BALANCE

2.1 TYPES OF GREENHOUSE GASES AND THE MAGNITUDE OF THEIR EFFECTS

The previous chapter discussed various possible techniques for cooling the earth or otherwise mitigating the effects of global warming. However, there are virtually no foreseeable easy, highly effective techniques for solving the problems without creating new environmental problems. Accordingly, let us now turn our attention to the reduction of greenhouse gas emissions. With regard to carbon dioxide, this chapter will examine only the behaviour of carbon dioxide at the earth's surface (i.e. the "carbon balance"); a detailed discussion of specific techniques for combatting the problem will be deferred to Chapter 3 and later.

Various gases exert a greenhouse effect (Fig. 2.1); the examples listed here are merely examples and the estimates for each gas vary widely. Figure 2.1 predicts the overall temperature increase between 1980 and 2030 at 1.5°C, and estimates that the contribution of carbon dioxide to this at around 0.7°C. The right-hand vertical axis indicates the warming effect of the gases relative to that of carbon dioxide (i.e. the cumulative contribution of the gases when the contribution of carbon dioxide is taken to be one). Since the contribution of chlorofluorocarbons (CFCs) and freon substitutes is half that of carbon dioxide, it can be assumed that these gases also have a major effect. The next main greenhouse gases are

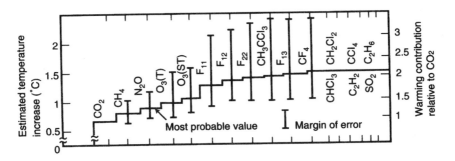

FIGURE 2.1 Effect on temperature of various greenhouse gases [Ramanathan *et al.*, 1985].

29

methane (CH_4), nitrous oxide (N_2O), and ozone (O_3). Although other gases also appear to have a greenhouse effect (e.g. various hydrocarbons, chlorocarbons and sulfur oxides), the effect is extremely small. Furthermore, sulfur compounds such as sulfur dioxide form aerosols, which block the sun's rays, thereby having a cooling effect on the earth; this makes it even more difficult to make predictions. The following sections will briefly examine the possibility of reducing the relative contributions of the principal greenhouse gases other than carbon dioxide.

2.2 THE NATURE OF GREENHOUSE GASES

The absorption of infra-red rays, which is the cause of global warming, primarily occurs by molecules comprised of three or more atoms; all the principal greenhouse gases (including carbon dioxide) fulfill this criterion. However, it must be noted that the greenhouse effect of a given substance cannot be determined merely from its characteristics or changes in concentrations. At a given wavelength, there is a logarithmic relationship between the change in light intensity (due to absorption) and the concentration. Therefore, if a substance is present at a higher concentration (or if another substance is present that absorbs radiation at the same wavelength) then the degree of the greenhouse effect per unit concentration decreases. Table 2.1 compares the absolute effects of various gases with that of carbon dioxide. Substances that have a considerable overall effect are not only those which have a pronounced ability to absorb radiation, but also those substances which currently are present in small quantities but which have an absorption wavelength different from other principal greenhouse gases (i.e. carbon dioxide and water vapor; see Fig. 1.10).

Another factor that must be considered is the life span of the substance in the atmosphere. For instance, even if a particular substance has a significant greenhouse effect, and is emitted in considerable quantities, the problem will not be serious as long as the substance decomposes quickly and in a manner whereby the greenhouse effect ceases. Although this effect is actually incorporated in the absolute evaluation, the matter is usually not debated to such depth; in most cases, the discussion just considers the magnitude of the effect based on the figures that are supplied.

Figure 2.1 and the last three rows of Table 2.1 give the contributions of each of the various gases. For example, Fig. 2.1 and Table 2.2 list the rate of contribution of carbon dioxide as approximately 50% or slightly less, which is the generally accepted figure. However, Yoshida and Murase (1990) give the value as 65%, and NEDO (New Energy and Industrial Technology Development Organization) (1992) gives the instantaneous value as 76%; when the effect is calculated over a period of 100 years or more, the contribution actually becomes 95%. The result

Table 2.1 Relative contribution to temperature increases of increased concentrations of various greenhouse gases.

	CO_2	CH_4	N_2O	O_3	CFCs	Source of data
Absolute effect relative to CO_2	1	20	100	2,000	10,000	(1)
Life span in atmosphere [years]	50–200	5–10	120	0.1–0.3	Approx. 100	(2)
Pre-industrial revolution concentration (ppm)	275	0.7	0.285	—	0	(2)
1985 level (ppm)	345	1.7	0.304	0.01–0.1	0.0006	(2)
Rate of increase in 1985 (%)	0.4	0.9	0.25	Approx. 1	5	(2)
Contribution (%)	65	15	3	5	>8	(1)
Instantaneous value in 1986	76.1	9.6	2.7		11.6	(3)
1986–2100	94.7	0.8	1.2		3.3	(3)

(1) Yoshida and Murase, 1990.
(2) Ramanathan et al., 1985, Dickinson and Cocerone, 1986.
(3) New Energy and Industrial Technology Development Organization, NEDO, 1992.
ppm = parts per million.

Table 2.2 Relative contributions of various sources of greenhouse gas emissions (OECD, 1992).

Emission source	CO_2	CH_4	N_2O	O_3	CFCs	Total
Energy use	35	4	4	6	—	49
Deforestation	10	4	—	—	—	14
Agriculture	3	8	2	—	—	13
Industry	2	—	—	2	20	24
Total	50	16	6	8	20	100

OECD = Organization for Economic Cooperation and Development.

varies greatly depending on the method of evaluation. The reason for differences in the instantaneous and long-term values, even when using figures from the same data source, is that the contribution of substances that undergo rapid decomposition (e.g. methane) becomes negligible over the long term. Nevertheless, as can be seen from the table, it is impossible to accurately attribute the life span of the substance. Furthermore, the evaluations will naturally be affected by projected changes in future emissions, although uncertainties also surround these. Rather than becoming involved in a deep debate about these, surely it is necessary to strive towards a reduction of the original emissions.

2.3 EMISSION SOURCES

Let us now consider the contributions of the various emission sources (Table 2.2). The emission source that immediately comes to mind is industry. However, the first point that needs to be understood is that half of the greenhouse gas emissions result from energy use. While industry does indeed use some of this energy, we must not forget that we also use a very large amount of energy in our daily lives. For example, the majority of the electricity that we use is produced at thermal power plants, from where carbon dioxide is discharged into the atmosphere. Agriculture is also a major source of greenhouse gas emissions. Fields and grazing land certainly do not exist naturally; also, the development of agricultural land is one of the causes of deforestation. A more detailed explanation of these emission sources will be given in later sections of this chapter.

Problems also arise when discussing the countries which are responsible for greenhouse emissions, since the results vary with the method of calculation (Figure 2.2). The question arises of which method should be used to evaluate these different sources—for example, should evaluation be performed at the site of emission of greenhouse gases during the import–export process, or be based

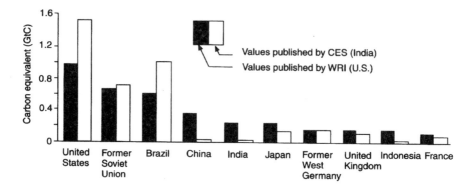

FIGURE 2.2 Greenhouse gas emissions by country [Asahi Shimbun, 1991]. CES = Center for Energy Studies, WRI = World Resource Institute, GtC = giga-tons of carbon.

on incomplete data such as that for deforestation etc. In addition, consideration must be given to the fact that political factors greatly influence such data. It is beyond doubt that the emissions from the United States and the former Soviet Union are considerable. But in the latter's case, data were not made public before the break-up of the Soviet Union, and since then the gathering of data has not been a high priority.

2.4   SOURCES OF FREON EMISSIONS AND THEIR SUPPRESSION

Freons* are known more for their role in depleting the ozone layer than as a greenhouse gas, and it is this that has already led to moves towards their total abolition. However, the fundamental problem lies in the fact that they are extremely stable substances with a life span of 100 years (Table 2.3). A further problem concerns so-called freon substitutes.** These substances, whose life span is short, tend to be chosen because they have only a slightly destructive effect on the ozone layer. As a result, most exert a small greenhouse effect (Table 2.3), although this does not hold true for all of them. [One example of a freon substitute is HCFC-22; this is a crude material of high molecular weight which is used as a coolant, but unfortunately it does produce damage to the ozone layer. Consequently studies are being performed to find possible substitutes for HCFC-22, but some of these have a greater greenhouse effect than HCFC-22]. There are additional problems with the switch from freons to alternative substances. First, there is a decrease in the efficiency of energy use. Second, additional energy is needed to achieve a reduction of emissions, and also for the recovery and decomposition of freons. These various factors lead to an increase in the total amount of energy required; accordingly, it is necessary to consider this as being an indirect increase in carbon dioxide emissions.

The production of all freon gases has ceased and they will no longer be used. However, a large amount of freons are lying dormant in air-conditioners and refrigerators used in the home and by businesses, and the same applies to

*Freons are chlorofluorocarbons, and are also referred to as CFCs. They are compounds which are composed of carbon, chlorine and fluorine, and are assigned numbers according to the sites at which chemical bonding occurs. Freons 11, 12, 113, 114 and 115 have a particularly destructive effect on the ozone layer. Besides being utilized as coolants, freons have been used in foaming agents, spray propellants, and detergents etc. Halons (i.e. materials in which part or all of the chlorine in the freon is replaced by bromine) are also used in fire extinguishers; as in the case of freons, three types have been designated as having a particularly harmful effect.

**Freon substitutes are substances that resemble freons and possess similar properties. The typical substances are HCFCs (hydrochlorofluorocarbons), in which some of the chlorine or fluorine is replaced by hydrogen, and HFCs (hydrofluorocarbons), which consist of carbon, hydrogen and fluorine. In the same manner as freons, the structure is indicated by the assignment of numbers.

Table 2.3 Greenhouse effect of freons and freon substitutes (*adapted from* Sekiya, 1990).

Freon substitute	Life span (years)	Greenhouse effect*
CFC-11 (CCl_3F)**	60	1
HCFC-123 ($CHCl_2CF_3$)	1.6	0.017–0.020
HCFC-141b (CH_3CCl_2F)	7.8	0.084–0.097
CFC-12 (CCl_2F_2)**	120	2.8–3.4
HFC-134a (CH_2FCF_3)	15.5	0.24–0.29
HFC-152a (CH_3CHF_2)	1.7	0.02–0.033
HCF-124 ($CHClFCF_3$)	6.6	0.09–0.10
HCFC-142b (CH_3CClF_2)	19.1	0.34–0.39
HCFC-22 ($CHClF_2$)**	15.3	0.32–0.37
HFC-125 (CHF_2CF_3)	21.8	0.51–0.65
HFC-143a (CH_3CF_3)	41	0.71–0.76

*Estimated relative value per unit mass.
**Substance to be substituted.

automobiles. It is therefore necessary to establish a system for the complete recovery of these substances when such items are discarded and to develop technology for achieving their subsequent decomposition. Indeed, a number of researchers are grappling with this latter question and such technology is being developed. If successful, this should lead to the amelioration and possible resolution of the problem. It could therefore be said that the problem lies in the very recovery of the substances.

2.5 SOURCES OF METHANE EMISSIONS AND THEIR SUPPRESSION

Next let us consider the sources of methane emissions. Methane is a major constituent of natural gas; it is also released following the anaerobic decomposition of living organisms, when something rots or decomposes in an oxygen-poor environment. However, since it is a gas which has been released in large quantities in the natural world since ancient times, it is difficult to ascertain the extent to which human activity has contributed to an increase in emissions. Furthermore, the gas which is released within nature undergoes natural decomposition, and in the past a balance was probably therefore maintained. In actual fact, an oxidation process exists in nature whereby the methane is converted as far as carbon dioxide, but the amounts involved are even more difficult to estimate. The estimated amounts of methane emissions from the various sources are shown in Fig. 2.3. The first thing to be noted, though, is that although much of the methane is released naturally, the increase in methane levels in recent years appears to be due to human beings. However, this is not necessarily due only to the direct effects of human activity (e.g. by industrial production); it could also be due to very slight changes in the natural environment (e.g. by livestock farming and rice cultivation).

As can be seen from the figure, although the estimates include errors, the combustion of biomass forms a source of methane emissions—even though in principle it does not lead to a net increase in carbon dioxide since the carbon dioxide was originally absorbed from the atmosphere. Also, natural gas, which is a low-carbon fuel, could also be termed a source of methane emissions due to releases that accompany its extraction and transportation. Wetlands are another major source of methane emissions; it thus seems perhaps a touch ironic that the protection of wetlands is called for from the standpoint of environmental conservation. The situation is extremely complex.

Since the majority of methane emissions are related to agriculture, countermeasures ought to be left to agricultural specialists. Several possibilities are becoming evident, although they are by no means simple. With advances in agricultural management it should be possible to cultivate rice that produces little methane. It is even said that the belching of cows is another source; a layman might say that it should be simple to develop medication that would prevent the belching, but it is not as easy as it sounds. In all cases, however, the problem has arisen from the increase in agricultural production that has accompanied the increase in population and the greater levels of wastage brought about by higher

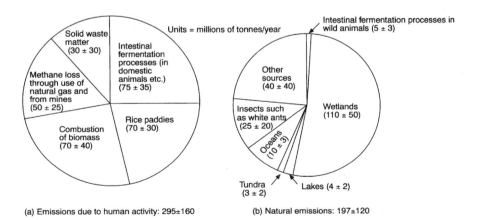

(a) Emissions due to human activity: 295±160   (b) Natural emissions: 197±120

FIGURE 2.3 Sources of methane emissions [*adapted from* OECD, 1992].

living standards. It might appear that if garbage were placed in landfills without being combusted, carbon dioxide emissions would be suppressed or at least delayed, but in actual fact this forms a major source of methane emissions.

In some places, methane is collected from waste landfills and used as an energy source; however, it is extremely difficult to collect all of the gas.

2.6   SOURCES OF EMISSIONS OF OTHER TRACE GASES, AND THEIR SUPPRESSION

Other gases which are present in small quantities and which have a greenhouse effect include nitrous oxide, ozone, trace hydrocarbons, sulfur oxides, and PAN (peroxyacyl nitrate).

Let us first consider nitrous oxide, (N_2O), also known as laughing gas. It was formerly employed as an anesthetic, but is now used as an aid for combustion due to its strong oxidant properties. It has also been implicated in the depletion of the ozone layer, although it has not been completely clarified whether its role is positive or negative.

Sources of nitrous oxide emissions are shown in Fig. 2.4. As is evident from the chemical formula (N_2O), it is related to the family of nitrogen oxides (NO_x), which include compounds such as nitrogen oxide (NO), a source of acid rain. For this reason, regulations concerning its production are strongly related to those stipulated for production of NO_x. If it is only the nitrogen present in the fuel that undergoes combustion to form NO_x, the latter is termed "fuel NO_x";

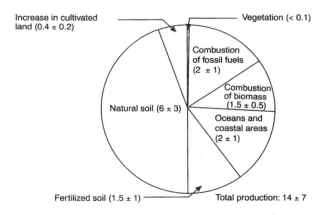

FIGURE 2.4   Sources of nitrous oxide emissions (in million tonnes/year) [*adapted from* Bouwman, 1990].

generally, however, when combustion takes place at temperatures greatly in excess of 1,000°C, the nitrogen in the air also combines with oxygen to form large quantities of NO_x (this is known as "thermal NO_x"). The production of thermal NO_x is greatly reduced if combustion takes place at temperatures below 1,000°C; however, nitrous oxide (N_2O) is produced at this time. Prior to nitrous oxide being identified as a potential problem, none of the combustion experts had foreseen this eventuality, and a method for monitoring its levels had not even been established. However, immediately after it was recognized as a possible problem it became a major topic of discussion at conferences on fluidized bed combustion (which is a technique of coal combustion). Many researchers energetically began the search for technology that would suppress the emission of nitrous oxide, and the very fact that such energy is being devoted to this goal makes it likely that a good solution will be found in the near future, although the situation has now been somewhat complicated by the realization that nitrous oxide also plays a positive environmental role by blocking the destruction of the ozone layer by freons.

Next let us consider the case of ozone. The ozone considered here is tropospheric ozone which exists close to the earth's surface and which (due to its short lifetime) is different from the stratospheric ozone that is being depleted by the release of freons. Certainly the problem related to stratospheric ozone levels (and in particular the creation of an ozone hole near the south pole) is extremely serious; on the other hand, from the point of view of global warming, ozone is a greenhouse gas which absorbs infra-red rays at wavelengths outside the range absorbed by carbon dioxide. Tropospheric ozone is produced from nitrogen oxides as the result of a photochemical reaction, and therefore the first countermeasures that come to mind are those techniques used to reduce the emissions of NO_x. Photochemical reactions also involve gaseous hydrocarbons such as methane, carbon monoxide (CO) and other gases present in minute quantities. In the future there are many points which must be investigated and clarified in relation to the mechanisms underlying ozone formation; the measurement of ozone concentrations at widespread locations; the estimation of its future concentrations; and the evaluation of the contribution of ozone to global warming.

Sulfur oxides (SO_x) are also known to have a greenhouse effect, but at the same time these lead to the formation of mist within the atmosphere and are thus said to lead to global cooling; many points relating to their effects are as yet unclarified. Emissions of sulfur oxides could be reduced by desulfurizing flue gases (alternatively, this could be done within the furnace).

Attention must also be paid to carbon monoxide since, although its effect is rather indirect, its behavior is strongly related to that of methane and ozone. Approaches to the reduction of carbon monoxide emissions would be to prevent incomplete combustion during industrial processes or when engines are used for transportation (in the same way as in the case of hydrocarbons such as methane).

At the present time, it seems that the measures aimed at reducing the concentrations of both sulfur oxides and carbon monoxide are being devised more from the point of view of pollution prevention rather than because of their greenhouse effect.

Carbon tetrachloride (CCl_4) is similar to freons, and moves have taken place to cease its production. Consequently, the amounts released in future are likely to decrease. Since hydrocarbons other than methane are only present in small quantities, their warming effect is not so large, but they are expected to have the same type of effect and countermeasures would have many points in common.

It is also possible that in the future various other new types of gases will be developed, as happened with freons. It is therefore necessary to adequately investigate their possible environmental effects before these become a problem.

2.7 CARBON DIOXIDE: ITS UNIQUENESS, AND FUTURE POLICY DIRECTIONS

Let us now focus on carbon dioxide, the main substance responsible for global warming. It is important to recognize at the outset that carbon dioxide is in many respects quite different from the other gases that produce global warming.

The gases discussed so far have various deleterious effects: freons lead to destruction of the ozone layer; NO_x and SO_x cause acid rain; and some greenhouse gases (such as freons, ozone, methane and nitrous oxide) have a considerable greenhouse effect despite being present in only small quantities. When substances are released in small quantities, whether they are substances which produce a warming effect or which cause an environmental problem, it should be possible to develop appropriate technological countermeasures. If we are prepared to sacrifice a small amount of energy, advances in engineering and agricultural technology should allow the recovery and removal of these substances; they could then additionally be converted into harmless but useful man-made materials; in industrial processes, it is likely that substitute materials could be used. While the current situation is indeed difficult, it is likely that such technology could be successfully developed (although, of course, the value of developing these countermeasures depends on the magnitude of the problem, which in turn depends on the individual substance).

The first difference between carbon dioxide and the other greenhouse gases is that it was present in the atmosphere in large quantities before the existence of human beings. As described in Section 1.8, carbon dioxide is absolutely vital for the maintenance of the earth's environment and in order to prevent the earth turning into a planet of ice. Without carbon dioxide, no vegetation would be able to participate in photosynthesis, and all animal life would become extinct.

The second difference is that, as described in this chapter and the next, carbon dioxide will inevitably be released as long as fossil fuels are combusted; and at present fossil fuels account for nearly 90% of energy use.

We must promote techniques that lead to the efficient use of energy, and in Japan's case it is necessary to transfer her advanced technology regarding energy conservation. However, even if this is done, carbon dioxide will continue to be emitted. Is it possible to switch to natural forms of energy? Biomass and other natural forms of energy were the main source of power in Japan during the Edo period (1603–1867), and hydraulic power fulfilled this role until after the Second World War. This was followed by a historic shift to fossil fuel energy; at the present time, the safety and environmental aspects of nuclear power are sources of great concern to the general public, which means that the era of fossil energy is likely to continue for some time. Needless to say, it is still necessary to continue the development of alternative forms of energy such as solar energy, and (provided the problem of assuaging public disquiet about the safety and waste disposal aspects can be overcome) also nuclear power.

The third difference between carbon dioxide and the other gases is that it cannot be artificially converted into carbon or organic compounds with the expenditure of only a small amount of energy. In practice the conversion into either harmless or useful organic materials requires enormous amounts of energy; indeed, the energy required is greater than that produced when carbon dioxide is released during the combustion of fossil fuels. Under such circumstances, there would be no rational need for using fossil fuels in the first place.

Among the scientists who advocate the effective use of carbon dioxide, there are some who take the position that carbon dioxide should be used as a raw material for the manufacture of chemicals. However, the amount of fossil fuels used for petrochemical products such as vinyl account for only a minuscule proportion of the total use of fossil fuels, and the use of such small quantities is surely permissible. Even if it is assumed that we must reduce carbon dioxide emissions to zero, another possibility (instead of using carbon dioxide as a raw material) would be to utilize biomass (for details, see Section 5.2).

One method of fixing carbon dioxide would be to first recover the gas, and then sequester it; the fixation could be achieved either in that form, or in the form of some other inorganic substance. Another approach would be to take advantage of a natural process of absorption and fixation (e.g. by utilizing vegetation or the oceans). Indeed, these strategies are the only other options.

The rest of the discussion will examine in greater detail the fundamental problems associated with carbon dioxide and with the countermeasures that could be adopted.

So as to avoid any possible confusion, however, it would be best to first briefly mention a few points. The author is considering that time at which the global warming problem has been acknowledged in all its extreme gravity, and when the enormous impact on human beings has been fully realized; nevertheless, it is hypothesized that at this time the alternative forms of renewable energy which will be described in Chapter 3 have not yet been developed, and that the energy conservation techniques discussed in Chapter 4 have not yet on their own proved adequate. It is at this time that most of the countermeasures mentioned in Chapters 5–7 will need to be adopted.

These are certainly important matters from the extremely long-term point of view; however, given the current situation in the world, it must be recognized that, rather than developing measures to combat the carbon dioxide problem, a greater problem concerns the curtailment of harmful substances which are responsible for regional pollution problems, such as NO_x and SO_x.

Countermeasures against carbon dioxide based on existing chemical technology (including the use of biological materials) are either non-existent or are extremely difficult to implement. Although the field has heretofore been referred to as chemistry, it is the belief of this author that this alone is inadequate, and that future research needs to be multi-disciplinary and have a clear aim in mind rather than abstract objectives.

As indicated before, it is essentially impossible to recover and use carbon dioxide efficiently as an organic material simply by the further development of existing techniques. Nevertheless, the government came out with the rosy "Action Plan for Mitigating Global Warming" which misleadingly makes people think that such techniques are in fact possible. Of course, as is shown in Fig. 2.5,

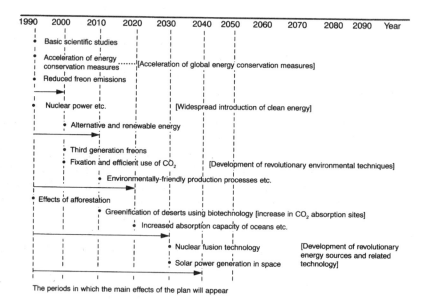

FIGURE 2.5 Action plan for mitigating global warming [Ministry of International Trade and industry, 1990].

several important topics have been identified by the government that do indeed need pursuing, such as afforestation (to prevent the destruction of tropical forests) and the development of alternative sources of renewable energy etc. These of course involve considerable difficulties with respect to policy considerations, the economy and technology, and these difficulties must be fully appreciated. The essential problem is that the government's programs include the projects mentioned below, the goals of which are essentially unattainable but which are simple to set in motion since they follow on directly from existing techniques, and are deemed appropriate for signalling to foreign countries what can only be described as a masquerade on the part of Japan to give the impression that she is pursuing research that is kind to the environment. However, the researchers themselves are caught up in the process, and in order to carry out research or obtain the funds for research, plans concerning research projects have to be submitted in accordance with government policy before a basic debate about the matter is conducted and before the matter is considered at a fundamental level. This state of affairs must be strongly criticized. The technological principles involved in the debate are discussed in detail in Chapter 5 and later chapters, but to summarize briefly, the fixation of carbon dioxide cannot be achieved by sequestration using catalysts, electrochemical fixation, fixation by coral or carbonaceous calcium, or by biotechnology techniques. The author definitely does not deny the value of such research. But if research is to be conducted, then the situation surrounding that research needs to be clearly understood. Surely we should commence studies which will enable us to correctly evaluate possible strategies for dealing with the carbon dioxide problem. As an illustration of this, numerous researchers who have recently begun studies on the fixation of carbon dioxide have essentially not actually studied this. In reality, the subject of the research may form a part of the development of an energy system that uses carbon dioxide as a medium. However, no real evaluation is conducted of the overall performance of the whole system that uses carbon dioxide as a primary energy carrier; rather, this is something which was not part of the original project but which had the label of "research objective" attached at a later date. This compromise demands investigation from the viewpoint of energy efficiency, but in this particular case it is merely acting as a substitute for the environmental problem.

What requires attention now is not the problem of how carbon is related to the energy system itself; in the final analysis the question to be addressed is how to construct a system of energy production and supply that has a high overall efficiency. In this respect, the question of the participation of carbon dioxide is irrelevant, and in order to solve the problems related to carbon dioxide we actually need for once to exclude consideration of carbon dioxide so as to avoid complexity.

There is another area in which the excuse of addressing the 'carbon dioxide problem' is used for promoting research that in reality aims at other objectives (such as the development of energy media themselves and the production of new materials)—and it is important to see this for what it really is. However, in the final analysis even the Japanese government, which should be taking the lead in looking at this problem, is not providing any guidance at all, but is merely adopting the posture of making substantial investments in research aimed at solving the carbon dioxide problem.

The energy problem itself is in essence the problem of carbon dioxide, and while humans use the existing precious fossil fuels efficiently and without waste, they must also establish an energy system based on alternative energy which does not use fossil fuels. The only way of fundamentally solving the carbon dioxide problem is to change the structure of the current system, which consumes vast quantities of energy. At the same time, it must also be recognized that an even more fundamental problem is that of overpopulation.

The above approach appears to be the only logical direction for us to take if we are to attain fundamental solutions to the problems directly concerned with carbon dioxide. It should be noted that this direction is the same as that indicated by the Club of Rome when they discussed the problem of energy resources. That is, notwithstanding the fact that the carbon dioxide problem should act as a great spur to the development of alternative forms of energy, present policies are not doing this, and this is a source of great dissatisfaction. This is not to say that technological measures for dealing with carbon dioxide itself are entirely absent, but the countermeasures should only be carried out as emergency measure, and when a serious problem involving climate change develops and cannot be resolved by any other means. From the standpoint of resources, these measures are ones which will be regretted later.* It is to be fervently hoped that such emergency measures will not be taken, but rather that other approaches (i.e. the development of alternative forms of energy) will be adopted.

Finally, let us consider the question of a carbon tax.** Although this is frequently debated (and in a number of countries has actually been introduced) the goal of reducing long-term carbon dioxide emissions could be achieved more effectively by the introduction of an energy tax. As will be explained in Chapter 3, if consideration is given to a shift between fossil fuels, then in the case of a switch from coal to natural gas (which releases less carbon dioxide than coal per unit of energy produced) unless alternative forms of energy have been developed

---

*"Regrettable" policies can be defined as those which squander resources on policies aimed at counteracting global warming, but which use even more of these resources (and hence priceless energy) than has been the case so far (as has happened in the case of carbon dioxide recovery). In the event of the perceived problem not actually becoming reality, this would mean that human beings would have simply wasted those resources. On the other hand, "non-regrettable" policies are those that will not be regretted if global warming does not in the end take place (or if its effect is minimal). To put it in concrete terms, the term refers to policies consistent with the direction in which the human race ought to move, i.e. energy conservation, conversion to alternative energy sources, and afforestation etc.

**It has been suggested that a tax be imposed directly in proportion to the amount of carbon contained in fossil fuels, since this carbon is the source of carbon dioxide emissions. Indeed, some countries have already implemented such a tax, although the tax is not levied on the carbon in natural forms of energy such as biomass. For details, see Chapter 8.

by the time the natural gas has been used up, we would have to convert back to the high-CO_2 emitting coal, and in the end the amount of carbon dioxide in the atmosphere would be the same. Furthermore, that would result in a less effective use of the natural capacity of the oceans to absorb carbon dioxide (for details, see Chapter 7) and could even lead to an increase in atmospheric carbon dioxide concentrations.

The important point is that a tax is needed to provide an incentive for greater efforts to be put into the search for technological means to improve energy conservation and to encourage the development of natural energy sources such as solar energy, geothermal energy, and tidal energy. Until such a time comes, it is extremely important to conserve forests and pursue afforestation, but in addition it is necessary for the Japanese government to make funds available to facilitate the introduction of highly-efficient equipment into countries which are far less energy-efficient than Japan.

2.8   THE GLOBAL CARBON BALANCE

Let us begin with something that is very obvious. That is, carbon dioxide does not escape into space. The reason that carbon dioxide does not escape into space is simply due to the effect of gravity. The molecular weight* of carbon dioxide is 44, whereas the molecular weights of oxygen and nitrogen are respectively 32 and 28. Carbon dioxide could therefore not escape into space until after the escape of the lighter gas of oxygen, which would mean that all human beings would first die of asphyxiation. If carbon dioxide were to dissociate and form carbon monoxide, the molecular weight would become 26, but that process would require a very large amount of energy and is unlikely to occur naturally. It can thus be safely concluded that carbon dioxide is not going to disappear into space.

Instead of vanishing into space, the world's carbon proceeds through a cycle during which it passes through various different forms. This cycle has continued since long before the advent of the human race. The changes occur extremely slowly during the carbon cycle, and the forms in which the carbon exists have gradually changed.

Figure 2.6 shows the carbon cycle at the earth's surface. Although the current partial pressure of carbon dioxide amounts to a mere 3.6×10^{-4} atmospheres, in the early stages of the earth's history it was equal to several dozen times present day atmospheric pressure. The natural power of rain and wind produced the

---

*This is the weight of approximately 6×10^{23} molecules (this is Avogadro's number, which is defined as the number of carbon atoms contained in 12 grams of the ^{12}C isotope). The greater the molecular weight of a gas, the greater its density.

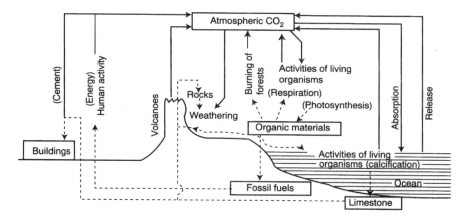

FIGURE 2.6 The carbon cycle at the earth's surface.

absorption of carbon dioxide by rock; subsequent weathering led to the carbon entering rivers, and finally being transported to the oceans (for details, see Chapter 5). It was then converted in the oceans by coral etc. into the form of carbonates (calcification); at this time, nearly half of the carbon dioxide absorbed by rock weathering was returned to the atmosphere but over half of it sank to the sea bed in the form of lime. Also, photosynthesis by living organisms resulted in some carbon dioxide being accumulated as fossil fuels. Such changes occurred over a period of a few billion years, finally producing the earth as we know it today.

Over a period of one hundred years, or even on a time scale of several tens of thousands of years, virtually none of the above changes in the balance can be observed. That is, the changes are almost infinitesimally small. Or rather they would be infinitesimally small (and the earth would be in a stable state even now) if it were not for the actions of human beings. The above changes are certainly taking place, albeit at a glacial pace with the continuation of weathering and the accumulation of carbonates on the sea bed. However, approximately the same quantity of carbon is being returned to the surface by changes in the earth's crust; also, carbon dioxide is being returned to the atmosphere due to volcanic action, and rock formation is again taking place. Approximately the same amount of carbon dioxide that is removed by vegetation during photosynthesis is transferred back to the atmosphere as a result of respiration by those very same plants and by the respiration of animals (including humans), and (on rare occasions) as the result of forest fires. These actions of course take place every single year. The oceans also release carbon dioxide into the atmosphere in equatorial regions, but this is balanced out by the absorption of virtually the same quantities in northern regions. (The processes involving rock, vegetation and the oceans are described in detail in Chapters 5–7).

As the seasons change, the processes repeat themselves and the overall change ought to be zero. However, the effects of human activity must also be added to the equation. This includes a range of activities that vary from the use of fossil fuels to the felling of trees and the production of cement. All of these lead to the release of carbon dioxide into the atmosphere, and it is this which has led to the carbon dioxide problem.

2.9 A MISSING SINK

This author recalls the words of a well-known researcher who stated that the use of fossil fuels* by humans released at most 5.5 gigatons of carbon dioxide. He continued by claiming that vegetative photosynthesis would remove in excess of 100 gigatons annually, and that the problem would therefore be resolved by the actions of plant life.

Under steady state conditions, Fig. 2.7 shows the amount of carbon dioxide as well as the amounts of annual exchanges between the land and the atmosphere, between the atmosphere and the oceans, and between the oceans and marine organisms. In discussions on vegetation and the oceans (which will be covered later), the figures are actually wildly inaccurate, with great variations in the estimates. Also, the total for fossil fuel deposits is only approximate, and may increase further if fossil fuels on the sea bed are also included. The amount of carbon that is transferred as a result of rock circulation has been omitted from the diagram since it is believed to be not so large; however, this author is unconvinced that this assumption is valid. At any rate, for the sake of argument let us assume in the following discussion that Fig. 2.7 is accurate.

Although all the figures are merely estimates, there is a vast amount of carbon in existence and even under steady state conditions an enormous amount of carbon is being transferred around the earth. The amounts are indeed colossal, and dwarf the figures of 5.5 gigatons for the annual release of carbon from fossil fuels and for the 3.5 gigatons remaining in the atmosphere.

But let us now recall what was stated earlier about a state of equilibrium. In a steady state there is no overall change, no matter how long the time frame. The global warming debate concerns a disruption of this steady state, even though the figures may seem rather insignificant. The argument over global warming is

---

*Besides fossil fuels, we also need to consider materials such as methane which are trapped deep underground and which may have been formed when the earth was first created (although this has not yet been confirmed). Furthermore, although the amount of carbon dioxide produced during the manufacture of cement is small compared with that emitted from fossil fuels, it cannot be discounted, and therefore should be included in the debate on fossil fuels.

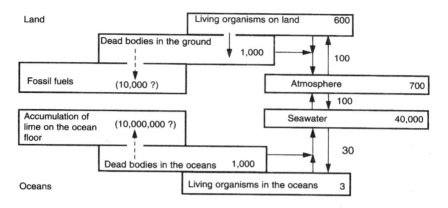

FIGURE 2.7 Amount of carbon on the earth's surface and amount exchanged per year under steady state conditions (in gigatons) (solid arrows = large exchange, dotted arrows = small exchange).

about a change of several degrees per century, or an annual change of less than 0.1°C. This seems small in comparison with even diurnal temperature fluctuations. The sun provides the earth with large quantities of energy but, even though the earth's temperature changes by over 10°C per day, winter forever remains as winter.

Perhaps the situation can be clarified by making an analogy with economics. Let us imagine 100 people with an annual income of $100,000 per year. Some people repay bank loans. Some people save in order to put down a deposit. If all these people saved an annual amount of $3,500 and worked for 40 years, they would have accumulated an average unused amount of $140,000 by the time of their deaths. However, such a situation is actually inconceivable. Some of the people would be fond of gambling, and some would waste their money. On top of that, they would use money in their old age. As a result, in the long term there would be a great variation in the amounts each individual would have, although the overall money in circulation would be constant when taking inflation into consideration. In exactly the same way, no matter how long vegetation exchanged an annual 100 gigatons of carbon with the atmosphere, it would be extremely difficult to squeeze out an annual average of 3.5 gigatons from that total.

The final conclusion of this chapter is previewed in Fig. 2.8. A state of non-equilibrium in the carbon cycle (or carbon balance) can be observed, which never occurred until the actions of human beings affected the process. Here again the natural transfer of carbon as a result of rock circulation is not large, and human activity is not likely to greatly affect this amount (although it must be said that the situation is not well understood); for this reason it has been omitted from the diagram, even though a possible increase in weathering processes (see later) means that there is a slight chance that the effect is noticeable.

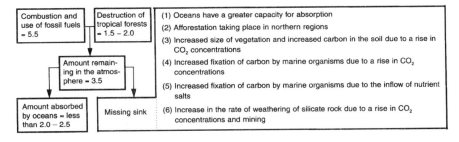

FIGURE 2.8   Non-steady state transfer of carbon resulting from human activity averaged over 1978–1988 (in gigatons per year).

The next implication of Fig. 2.8 is that not all the carbon dioxide which is released from fossil fuels remains in the atmosphere (in addition, deforestation releases an amount of carbon which is equivalent to approximately one-third of the carbon dioxide released from fossil fuels). The quantity of carbon dioxide remaining in the atmosphere amounts to only about half of the quantity released. But if it is assumed that it cannot escape into space, the question remains of what happens to it. A portion is certainly absorbed by the oceans, but the estimated quantity absorbed is very small. So where does it go? The carbon which is unaccounted for amounts to more than 1–2 gigatons per year.

2.10   THE CONTINUING INCREASE IN ATMOSPHERIC CARBON LEVELS

Estimates of past changes in atmospheric carbon levels can be calculated from the concentrations of gas trapped in Antarctic ice (changes over the last 200 years are shown in Fig. 2.9). With regard to more recent changes, the direct

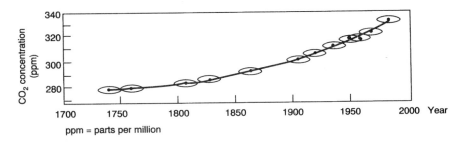

FIGURE 2.9   Changes in atmospheric CO_2 concentrations in the Antarctic over the past 200 years [Neftel *et al.*, 1990].

measurement of carbon levels by Keeling at Mauna Loa in Hawaii are well
known. His observations commenced before carbon dioxide became a problem
and have continued over several decades; they are the most accurate and most
quoted data available (Fig. 2.10).

The recent annual average data on carbon dioxide levels collected at Mauna
Loa are shown in Table 2.4. In the 1960s, the annual increase was of the order of
1 ppm (part per million), but recently that figure has risen to almost 2 ppm. The
average annual increase in carbon dioxide concentrations between 1978 and
1988 was 1.6 ppm. If this is multiplied by the total number of moles of gas
present in the atmosphere (approximately 1.8×10^{20} mol), and then multiplied
again by the atomic weight of carbon (12 g/mol), this represents an annual

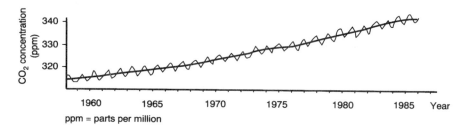

ppm = parts per million

FIGURE 2.10   Changes in CO_2 concentrations at Mauna Loa, Hawaii (at latitude
20°) [Keeling and Wharf, 1990; Keeling *et al.*, 1995].

Table 2.4 CO_2 emissions from fossil fuels and the amounts remaining
in the atmosphere (adapted from Marland, 1990).

Year	CO_2 concentration (ppmv)	Amount of CO_2 released from fossil fuels etc. (GtC/yr)			
		Total	Solid fuel	Liquid fuel	Others*
1963	319.1	2.86	1.40	1.05	0.41
1968	322.8	3.60	1.46	1.55	0.59
1973	329.8	4.65	1.59	2.24	0.82
1978	335.2	5.08	1.80	2.38	0.90
1983**	342.8	5.08	2.00	2.16	0.92
1988	351.2	5.89	2.39	2.39	1.11

* The majority is gaseous fuel, but there has also been a large increase in the
release from cement (from 0.05 GtC in 1963 to 0.15 Gtc in 1988).
** The use of liquid fuel decreased greatly following the oil shock.
ppmv = parts per million by volume.
GtC = gigatons of carbon.

accumulation equal to approximately 3.5 gigatons of carbon equivalent. This figure is believed to be highly accurate.

Now let us look again at Fig. 2.10. Due to the changes of the seasons, a maximum is observed around May and a minimum around September to October. It is said that the reduction in carbon dioxide concentrations is caused by increased photosynthesis of both land and marine vegetation, and that the increase in levels is due to the decomposition of organisms and the increased use of fossil fuels in winter. However, is this explanation correct?

In recent years atmospheric carbon dioxide concentrations have been measured at numerous locations (Fig. 2.11). The southern hemisphere clearly exhibits a maximum around October. However, the changes observed in the southern hemisphere, which contains a large proportion of water, are small in comparison with those seen in the northern hemisphere. From this, it can be deduced that the land has a considerable influence on the changes in carbon dioxide levels. Furthermore, recent observations of the concentrations of carbon isotopes have added support to the theory that this change is due not to fossil fuels but to vegetation. This seems feasible considering that whereas the amount of carbon being released from fossil fuels is of the order of only 5.5 gigatons per year, the quantity involved in exchanges with the atmosphere amounts to 100 gigatons.

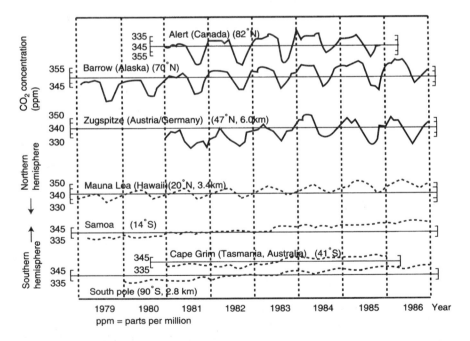

FIGURE 2.11 Changes in CO_2 concentrations at various locations on the earth's surface [Okamoto, 1989; Source: Meteorological Agency].

Close inspection of Figs. 2.10 and 2.11 reveals that each year the change in concentration is becoming greater, even though the annual increments are extremely small. In fact, this increase rivals in size the increase in total fossil fuel use. It is difficult to assume that all of this increased use of fossil fuel occurs only in winter. In that case, it presumably means that the photosynthetic capacity of vegetation is rising. However, even if this photosynthetic capacity did rise, this merely means that the amounts involved in the carbon cycle depicted in Fig. 2.7 have risen. It does not provide any definitive proof that it is a cause of a missing sink in Fig. 2.8. Of course, according to the syllogism it might be appropriate to consider that the amount of vegetation itself had increased; this may indeed be a possibility, but at the moment insufficient evidence is available for reaching a conclusion.

Let us now analyze the behavior of the carbon dioxide that is released by examining the worldwide distribution of carbon dioxide concentrations. First, Fig. 2.12(a) shows observed levels over a three-year period at various latitudes. Clearly, the concentrations in the northern hemisphere are higher than those recorded in the southern hemisphere because of the effect of fossil fuels. As previously stated, there are greater annual changes due to vegetation in the northern hemisphere, but this is not a cause of the higher concentrations of carbon dioxide because under steady state conditions the total amount of carbon dioxide that is absorbed is equal to the total amount released, i.e. there is no overall change.

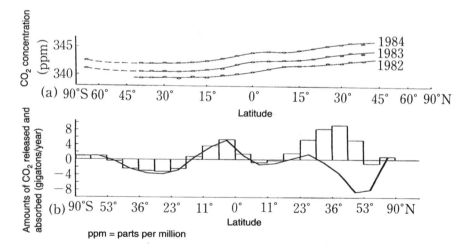

FIGURE 2.12 (a) Average latitudinal distribution of CO_2 concentrations at the earth's surface, and (b) the estimated latitudinal distribution of CO_2 emissions and absorption [Tanaka, 1991]. (b) The data in the bar graph include emissions from fossil fuels; the irregular line indicates data values after fossil fuel emissions have been subtracted.

This form of reasoning is rather intuitive. However, let us now analyze the situation rather more rigorously using the atmospheric transport model (i.e. a model of material transport in the atmosphere). That is, using current knowledge concerning atmospheric movements, it is necessary to account for such a distribution by investigating where the carbon dioxide is released from, and where it is absorbed. The data in the bar graph section of Fig. 2.12(b) confirm that emissions in the northern hemisphere are greater due to the combustion of fossil fuels.

As will be described in the next section, it is easy to estimate the proportion of these emissions that is derived from fossil fuels; if this amount is then subtracted from the total, it yields an irregular line of the type shown in Fig. 2.12(b). It is thought that the release in equatorial regions and absorption in high latitudes is due to circulation of the oceans (this will be discussed further in Chapter 7). The gist of the argument is that if the amount of emissions due to fossil fuels is subtracted, then it appears as if more carbon dioxide is being absorbed in the northern hemisphere, which may be due to the previously-mentioned increase in vegetation. However, as was stated earlier, the results of such an analysis include the effects of the oceans. In order to come to a clear conclusion that one cause of a missing sink is indeed an increase in the total amount of vegetation, more evidence is required.

2.11 RELEASE OF CARBON DIOXIDE FROM FOSSIL FUELS

Based on United Nations energy statistics, various researchers (e.g. Marland, 1988) have estimated the amount of carbon dioxide emitted worldwide from fossil fuels (including emissions arising from cement production). In short, this involves multiplying the production or consumption of each type of fuel by the amount of carbon contained within it. Coal has the greatest carbon content per unit of heat generated, followed by oil; natural gas has the least carbon per unit of heat generation, about half the value of coal (Table 2.4). The use of oil expanded greatly between 1960 and the early 1970s, after which came the increased use of natural gas etc. (included in the table under the heading "Others") and the restoration of coal's role. The emission figures are shown for 5-year intervals, but the average for the period 1979–1988 was approximately 5.5 gigatons, although this value does contain a certain degree of error. For example, the total also includes the amount buried underground, as well as the quantity preserved in the form of plastic etc. which is not burnt after it has been used. However, the great majority of fossil fuels are used as a source of energy, and most of that which is used as a raw material for the chemical industry undergoes either combustion or decomposition. At any rate, it is only a minuscule amount that does not end up being converted into carbon dioxide. The amount

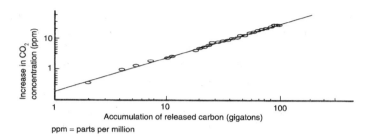

ppm = parts per million

FIGURE 2.13 Relationship between increased CO_2 concentrations at Mauna Loa, Hawaii, and total CO_2 released from fossil fuels worldwide since 1958 [Boden *et al.*, 1990].

which is used for non-commercial purposes was excluded from the statistics, but since the majority of the fossil fuels are used commercially, the statistics can be regarded as fairly reliable.

The time periods covered are almost exactly the same as those used to construct Table 2.4. This leads to the conclusion that approximately 60% of the carbon dioxide released from fossil fuels remains in the atmosphere.

As mentioned previously, observations of carbon dioxide concentrations have continued at Mauna Loa in Hawaii since 1958. Figure 2.13 shows the relationship between the increase in carbon dioxide concentrations observed since 1958 and the total amount released from fossil fuels during the same period. This diagram can also be interpreted as showing that 60% of the carbon dioxide released from fossil fuels since 1958 is still present in the atmosphere.

2.12 CARBON DIOXIDE RELEASED BY DEFORESTATION

The tropical rainforests have the important role of being a storage site for carbon at the earth's surface. However, they are being destroyed. The United Nations' Food and Agricultural Organization (FAO) has estimated that destruction of the tropical forests proceeded at an annual rate of 11.3 million hectares per year over the period 1981 to 1985 (Table 2.5). (This is roughly the same period as that covered in Table 2.4.) In contrast, the area undergoing afforestation in the tropics is only 1.1 million hectares per year, thus leaving an annual shortfall of 10.2 million hectares. This will be discussed in detail in Chapter 6, but there follows a very brief summary. In 1978, Woodwell *et al.* reported that the difference between the amount of carbon in living organisms per hectare of tropical forest is 170 tonnes/hectare more than the figure for carbon in living

Table 2.5 Average annual areas of deforestation and afforestation in tropical forests, over a five-year period (1981–1985) (adapted from Lanly, 1982).

Region	Area of tropical forests ($\times 10^6$ ha)	Area of tropical forest lost ($\times 10^6$ ha)	Area of tropical forest lost (%)	Proportion lost due to slash-and-burn agriculture (%)	Other Main causes	Area of afforestation ($\times 10^6$ ha)	Ratio of deforested land to afforested land
Tropical Asia	336	2.02	0.60	49	Immigrant farmers	0.44	5:1
Tropical Africa	707	3.68	0.52	70	Permanent agriculture	0.13	29:1
Tropical America	891	5.61	0.63	35	Overgrazing and immigration	0.54	11:1
Total	1934	11.30	0.58	49		1.10	10:1

organisms in grasslands.* The product of this value and the figure for the short-fall of 10.2 million hectares/year yields a value of 1.7 gigatons of carbon per year. However, considering the apparent inaccuracy of this figure, the author has estimated that the annual amount of carbon dioxide released into the atmosphere due to the destruction of tropical forests and creation of grasslands falls in the range of 1.5–2.0 gigatons (Fig. 2.8).

These results mean that less than half of the carbon dioxide that is released into the atmosphere actually remains there. If it is assumed that the contribution of land vegetation to the carbon balance amounts only to this, then it follows that the oceans must absorb over half the amount released into the atmosphere.

Let us now consider why these figures are inaccurate. Numerous people have calculated the amount of carbon dioxide released as a result of deforestation (Table 2.6). In 1977 Woodwell and Houghton estimated that the maximum

Table 2.6 Estimates of carbon released due to deforestation (Oikawa, 1990).

Study (and year)	World total (GtC/year)	Tropical area only (GtC/year)
Woodwell and Houghton (1977)	2.5–20.0	
Adamus et al. (1977)	0.4–4.0	
Bolin (1977)	0.4–1.6	
Brunig (1977)	6.0	
Wong (1978)	1.9	
Woodwell et al. (1978)	4.0–8.0	
Weiler and Crutzen (1980)	−2.0–2.0	
Brown and Lugo (1981)	−1.0–0.5	
Moore et al. (1981)	2.2–4.7	1.8–3.8
Olson (1982)	0.5–2.0	
Houghton et al. (1983)	1.8–4.7	1.3–4.2
Detwiler et al. (1985)		1.0–1.5
Yoda (1985)	1.8	
Houghton et al. (1987)	1.0–2.6	0.9–2.5
Detwiler et al. (1988)	2.0	
Woodwell (1989)		0.4–1.6

Note: A negative number essentially indicates absorption due to afforestation etc.

GtC = gigatons of carbon.

---

*This figure may become slightly higher. This is because data suggests that the area of deforestation may be larger (this will be discussed later); because carbon is lost from the soil; and because some rainforest may be converted into desert. On the other hand, data reported by Yoda (1982) suggests that the figure is smaller (for details, see Chapter 6).

global amount of carbon equivalent released was 20 gigatons. In the following year this maximum figure was revised downwards, but even so the figure of 4.0–8.0 gigatons was again proposed by Woodwell and his co-workers. In the early 1980s some estimates even suggested a negative total, implying the possibility that forests were bringing about a net absorption of carbon dioxide. However, the great discrepancy between the various estimates has meant that they can not be relied upon.

There is another source of worry. Many readers will perhaps have seen the photographs taken of the Amazon region by the man-made satellite Landsat. As the Amazon basin has gradually been opened up, deforestation has increased, and recent estimates of deforestation have been far higher than those published by the FAO; an example is shown in Fig. 2.14. Although there is a difference with countries such as Thailand which are bringing the felling of trees to a halt (or perhaps this is simply because the number of trees that can be felled has become smaller), the present estimates of deforestation in Brazil and India far exceed those of the FAO. Indeed, the FAO itself recently revised their estimate of 11.3 million hectares per year upwards to 16.9 million hectares (FAO, 1992).

Figure 2.15 shows the figures since 1820 for both the amount of carbon that has been lost from the soil and forests as a result of the felling of trees and which has then been released into the atmosphere as carbon dioxide, and also the amount released by the combustion of fossil fuels. Although fossil fuels are the primary cause of recent carbon dioxide releases, historically the main source has been release from forests. Before the industrial revolution, forests were converted into agricultural land, mainly in the developed nations such as European countries. This change lay behind the development of the advanced nations, and later these countries came to rely on fossil fuels; their wasteful style of living still

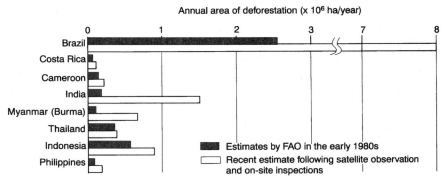

FIGURE 2.14   Recent estimates of areas of deforestation [*adapted from* Yoda, 1990].

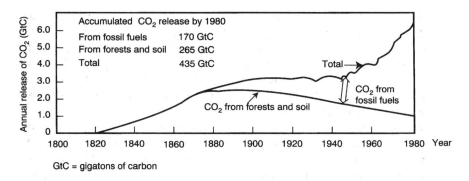

GtC = gigatons of carbon

FIGURE 2.15 Changes in the amounts of CO_2 released by combustion of fossil fuels and by the destruction of forests and changes in land use (forest destruction was estimated from measurements of ^{13}C levels in trees and from estimates based on the growth rings of trees) [Scientific Committee on Problems of the Environment (SCOPE) 1986].

continues today. Given these circumstances, one might ask whether the advanced nations really have the right to tell developing countries to stop felling trees without offering them compensation, just because carbon dioxide has become a problem.

2.13 THEORETICAL MODELS FOR THE ABSORPTION OF CARBON DIOXIDE BY THE OCEANS

Let us again look at the earth's carbon balance shown in Fig. 2.8. The questions of fossil fuel use and the destruction of tropical forests have been discussed in detail. If this represents the total of all sources of release and absorption between the land and the atmosphere, then the remainder must surely be accounted for by the oceans. From the equilibrium between the oceans and the atmosphere, and considering that the amount of carbon dioxide in the atmosphere is increasing, it would appear that the oceans are definitely acting as a sink. Unfortunately it has not been possible to perform measurements with a precision sufficient enough to allow a definitive conclusion to be reached.

The reasons for the lack of precise data stem from the techniques which must be employed for estimating the amounts of carbon dioxide absorbed and released. The first method is to measure the composition of sea water at the surface of the various oceans. If the amount of carbon dioxide dissolved in the sea water were greater than that which would be present at equilibrium with the atmosphere, then carbon dioxide would be transferred from the oceans to the atmosphere; if the opposite were observed, carbon dioxide would be absorbed.*

Unfortunately, it is not possible to perform the ocean/atmosphere exchange calculations at all locations. However, the partial pressure difference and the mass transfer coefficient (an indicator of the ease of mass transfer—in this case, the ease with which carbon dioxide can be transferred) both vary with the region and the season; the situation is further complicated by the fact that even for the same partial pressure difference, the rate of exchange will be high if the sea is very rough, whereas it will be low on a calm day. The difference between the partial pressure of carbon dioxide in the atmosphere and the average partial pressure due to carbon dioxide in the oceans (determined in terms of ionic concentration) has been accurately measured. The amount of carbon dioxide exchanged (obtained from the product of the mass transfer coefficient and the partial pressure difference) is integrated over the whole world to provide a global total. But because the mass transfer coefficient cannot be determined as accurately as the partial pressure difference, values are only approximate. Such calculations yield roughly equal values both for the amount transferred from the oceans to the atmosphere and vice versa, with the annual transfer amounting to approximately 100 gigatons (Fig. 2.7). It would have been desirable to incorporate the difference between the two totals into Fig. 2.8, but it would be impossible to accurately compute the difference between such large and almost equal figures when both figures contain margins of error.

Consequently, we should consider theoretical models for computing the amount of carbon dioxide absorbed by the oceans. Full details will be presented in Chapter 7, but a brief summary follows. The simplest model is known as the "Two Box" model (Fig. 2.16). This treats the oceans as consisting of two distinctly separate parts because the well-mixed surface layer and the well-mixed deep ocean are separated by a region of poor mixing (known as the "thermocline").

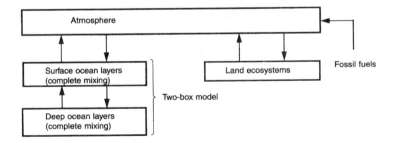

FIGURE 2.16 The two-box model [*adapted from* Inaba, 1990].

---

*Figure 7.6 shows the difference between this value converted into the concentration and partial pressure at equilibrium, and the partial pressure of carbon dioxide in the atmosphere.

The mixing that occurs between these two parts (or 'boxes') can be determined from the distribution of the concentrations of tracer substances such as isotopes of carbon. Instead of treating the deep ocean as a box, some other models consider the transfer of carbon dioxide by diffusion, and others are based on a bypass system with flows between the two boxes.

The results obtained from the various models allow a calculation of the amount of carbon dioxide absorbed by the oceans (Fig. 2.8). Simulations by oceanologists have suggested that the maximum absorption is only 2.0–2.5 gigatons of carbon equivalent, with the remainder treated as being as yet undetermined missing sinks. The term "maximum" was used here but this is because (as was mentioned in the previous section) it is expected that the ocean actually absorbs more carbon dioxide; efforts are being expended to therefore revise the models in such a way as to account for the increased absorption. Even if the models are modified in such a manner, the maximum absorption of carbon dioxide is only given as 2.0–2.5 gigatons because it is impossible to ignore the numerous measurements which have been reported for the distribution of isotopic tracers. Such amendments are likely to lead to an apparently convenient, but unsatisfactory model which infers a good transfer of carbon dioxide but a poor transfer of isotope tracers.

2.14   A MISSING SINK: JUST HOW LITTLE
WE KNOW ABOUT THE EARTH

Based on data for the period 1978–1988, Fig. 2.8 shows the most reliable figures for the current situation. The following discussion will consider the six possibilities for the existence of a missing sink.

As previously mentioned, the first possibility is the case in which the absorptive capacity of the oceans is far in excess of that predicted by theoretical models. The model described in the previous section is in the final analysis only a model, and various possibilities remain. Just because isotope tracers are used to determine the parameters in the model of concentration distribution, it does not necessarily follow that there is perfect agreement between the values which are predicted by that model and the actually observed distribution of tracers. Furthermore, tracer measurements are affected by nuclear testing (which increases the amount of isotopes in the environment), which in many cases has made it impossible to perform accurate measurements.

Kanamori et al. (1987) estimated the absorption by the oceans to date based on measurements of inorganic carbon in the sea water of the north Pacific Ocean. It must be noted, however, that in addition to the amount needed to bring about equilibrium with concentrations in the atmosphere, the inorganic carbon concentration in sea water also includes carbon that was introduced by the decomposition of living organisms. The decomposition of living organisms also produces nutrient elements such as phosphorus, whose concentrations in turn determine the amount of inorganic carbon that originates from living organisms. Figure 2.17 shows the results obtained by subtracting this amount and recalculating the atmospheric concentration for a state of equilibrium. Such calculations show that the concentration at depths of more than 1,000 meters corresponds to an atmospheric concentration of 250 ppm, which is slightly less than pre-industrial revolution levels; this implies that there is a slight resistance to exchange of materials with the atmosphere. It must be noted that the concentration increases linearly between a depth of 1,000 meters and the layers adjacent to the surface, and that therefore the inorganic carbon present in this region is carbon that appears to have been absorbed by the ocean. Of course, this result may not apply to all regions of the oceans, but this quantity is cited as a possible explanation of a missing sink.

The next possibility for a missing sink is related to the afforestation that is taking place now in northern regions (i.e. the former Soviet Union and Eastern Europe) and that which has occurred in the past. Although figures vary depending on the data used, it appears that the forested area is increasing, but even since the dismantling of the Soviet Union there has definitely been a shortage of information (information was sparse in the Soviet era because of government controls; after the collapse of the Soviet system, it might appear that much more information is available, but it is still insufficient). Furthermore, although not so much a question of the past as a possibility concerning future events, it is the northern regions that will become warmer as a result of global warming, raising the prospect that the boundary between the tundra and the northernmost treeline will be shifted further to the north. On the other hand, it is possible that the thawing of the frozen ground will lead to problems such as the release of methane.

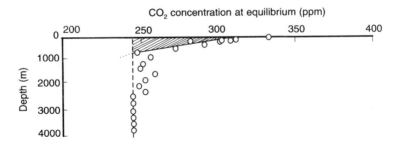

FIGURE 2.17 The concentration of atmospheric carbon dioxide calculated as an equivalent for the physically absorbed inorganic carbon concentration (i.e. adjusted for biogenic inorganic carbon concentrations) for various depths of sea water in the north Pacific Ocean [*adapted from* Kanamori and Ikegami, 1987].

The third candidate for the missing sink shown in Fig. 2.8 is that the amount of vegetation has increased as a result of an increase in carbon dioxide concentrations. In Japan, the physique of the Japanese people has improved dramatically since the end of the Second World War, which is attributable to the increased intake of nutrients. It must be stressed that the same point applies to vegetation. In actual fact there has been experimental verification that the carbon dioxide-rich atmosphere resulting from human activity has accelerated tree growth. Figure 2.18 shows calculated amounts of vegetation for increased carbon dioxide concentrations. It is predicted that the surface areas of leaves will increase, and despite the fact that light will weaken the total amount of vegetation will increase.

Figure 2.18 also predicts an increase in the carbon contained within the soil. This seems to be because the increase in photosynthesis has been accompanied by a rise in the quantity of organic materials falling on to the soil, and because the rate of accumulation has exceeded the rate of decomposition. With the greater carbon dioxide concentration there has been an increase in the rate of photosynthesis, which has in turn increased the production of leaves etc. and hence the supply of carbon to the soil. A possible secondary effect is that in future leaves will contain less nitrogen than carbon; however, this is mere conjecture. Such leaves will decompose in the soil less easily than leaves do today, with the result that this mechanism will increase the carbon content of the soil. It must be stressed that the above points are merely conjecture, and that there is no evidence that these form a missing sink; in addition, these views have been refuted by certain researchers.

Let us now consider the role of marine microorganisms. This will be dealt with in greater detail in Chapter 7, but in general the ratio (in moles) of carbon to nitrogen and phosphorus in ocean plankton is believed to be $106 : 16 : 1$ (the so-called Redfield ratio). Since inorganic carbon undergoes substantial dissolution in sea water, it is supposed that nitrogen and phosphorus are the elements

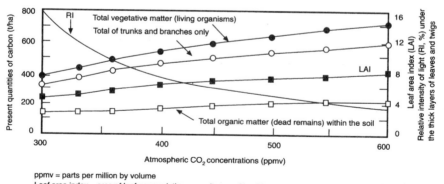

ppmv = parts per million by volume
Leaf area index = area of leaf accumulation per unit area of land (expressed as a percentage)

FIGURE 2.18   Predicted reliance of the organic matter content in forest ecosystems on CO_2 concentrations [Oikawa, 1989].

that are in short supply. There is in particular a severe shortage of nitrogen; using phosphorus as a basis for comparison, only 94% of the required nitrogen is available. On the other hand, the amount of carbon present is 9.6 times that which is needed (Redfield *et al.*, 1963). It is therefore widely recognized that even if the carbon dioxide concentration rises, there will be no increase either in photosynthesis or in the amount of carbon originating from marine life.

Nevertheless, a recent article in *Nature* (Riebesell, 1993) claimed that a rise in the atmospheric levels of carbon dioxide would cause a large increase in the amount of plankton produced. It is a theory which holds that carbon may become a limiting (or controlling) element, arguing that it is only possible to utilize the molecular carbon dioxide which is dissolved in the sea water (which amounts to less than only 1% of the dissolved carbon). Riebesell suggested that the conversion of bicarbonate ions to dissolved carbon dioxide molecules takes place only very slowly in sea water, but questions surround both the accuracy of this view, and also about whether the increased rate of photosynthesis is actually linked to an increase of living organisms in the world's oceans.

If it is assumed that, as mentioned above, the availability of both phosphorus and nitrogen is restricted, then one possible cause of a missing sink would be an inflow of nutrients. Calculations using the previously mentioned Redfield ratio show that if the entire annual production of fertilizer were to flow into the sea, then the annual amount of carbon undergoing fixation would be several hundred million tonnes. It is thus necessary in future to clarify the flows of nitrogen and phosphorus on a global scale. Also, it is ironic to note that the destruction of land ecosystems produces an outflow of nutrient-rich soil which may be a cause of a missing sink.

The final possibility for a missing sink is the weathering of rock. This is due to the absorption of carbon dioxide by silicates in rock and is part of the age-old reaction in response to the carbon dioxide (which existed in the earth's atmosphere under a pressure of several atmospheres) being forced into the earth. However, this reaction is not rapid enough to warrant evaluation, and is not likely to have much of an effect even if carbon dioxide concentrations rise. In contrast, the scale of human quarrying activities and soil erosion etc. has recently become so large as to pose an environmental problem, leaving the definite possibility that they may have an effect on the overall problem. However, if this is the case, that would mean that environmental destruction is becoming a source of carbon dioxide absorption, which would indeed be a source of irony.

2.15   THE SCALE OF THE PROBLEM

As stated above, estimates relating to the earth's carbon balance are by no means reliable, and the data shown in Fig. 2.8 are of doubtful accuracy. Thus while bearing in mind the slight doubt about the diagram's reliability, let us consider ways of resolving the carbon dioxide problem. Of course, some slight changes may have already been observed since the time the data in Fig. 2.8 was collected, and further changes may also be anticipated. However, this will not be discussed here. Nevertheless, there is one very important point that must not be forgotten, i.e. the missing sink(s). The question of whether the missing sink(s) can be relied on to solve the carbon dioxide problem depends on the particular mechanism involved. Needless to say, if the mixing sink(s) should disappear tomorrow, the premises for the following argument will be wrong. In order to solve the carbon dioxide problem, it is first necessary to more accurately estimate the global balance (including the missing sink). The following discussion is therefore based on the assumption that the missing sink(s) will continue to absorb carbon dioxide in the same manner as now.

The problem surrounding carbon dioxide concerns the amount remaining in the atmosphere, and thus it is necessary to either annually reduce carbon emissions to the extent of some 3.5 gigatons of carbon equivalent, or else remove that amount from the atmosphere. Since current emissions from fossil fuels amount to 5.5 gigatons, this would require a reduction of over 60%. The amount of carbon dioxide removed from the atmosphere by both the oceans and missing sinks is more than the amount remaining in the atmosphere; it would thus suffice if the absorptive capacity could be almost doubled.

Even a complete halt to the destruction of tropical forest would only halve the amount of carbon dioxide remaining in the atmosphere. Nevertheless, this is extremely important because the destruction of tropical forests is not a problem concerning only carbon dioxide, and also because it is definitely a feasible proposition. In order for it to be realized, though, it would be necessary to construct appropriate political and economic systems, and to also provide alternative energy sources for replacing the use of firewood in developing countries (which is one of the reasons for the continuing deforestation).

Atmospheric carbon dioxide levels would no longer increase if deforestation ceased, if the residual amount in the atmosphere were halved (as described earlier), and if in addition afforestation proceeded at the same rate as deforestation has so far occurred. Alternatively, the carbon dioxide problem could also apparently be solved if, in addition to the prevention of deforestation, either the emissions from fossil fuels were also suppressed by 30%, or else if the absorptive capacity of the oceans and missing sinks were raised by 1.5 times.

The question that then remains is whether any of the techniques under discussion are really feasible. That is, whether the goal can be achieved by only applying policies that will not be regretted later (using 'regret' in the sense defined in Section 7 of this chapter). Subsequent chapters of the book will examine the above countermeasures in greater detail and assess the chances of their being implemented.

References

Asahi Shimbun (Japanese newspaper). 1991, 25 Feb. Cited in Tomita, A. 1991. In Tomita, A. (ed.). Coal Utilization from the view point of Carbon Dioxide Problem. Report of Grant-in-Aid for Scientific Research, The Ministry of Education Science and Culture, Japan (in Japanese)

Boden, T. A., P. Kanciruk, M. P. Farrel (eds.). 1990. TRENDS'90, A Compendium of Data on Global Change. Carbon Dioxide Information Analysis Center, Oak Ridge National Laboratory, Oak Ridge, Tennessee, USA, 37831-6335. Cited in Tanaka, M. 1991. In Organizing Committee of 5th Symposium on University and Science (ed.). *Science of Global Environmental Change.* p. 169, Kubapuro Pub., Tokyo, Japan.

Bouwman, A. F. 1990. *Soil and the Greenhouse Effect.* John Wiley & Sons, New York, USA. Cited in Kawashima, H. 1993. In Kojima, T. (ed.). Seminar on Global Environment, Vol. 5, p. 38, Ohm Pub., Tokyo, Japan (in Japanese)

Dickinson, R. E., R. J. Cicerone. 1986. *Nature,* **319**, 109

Inaba, A. 1990. *Kagaku Kogaku* **54**, 18 (in Japanese)

Kanamori, S., H. Ikegami. 1987. *Meteorological Research Notes* **160**, 147 (in Japanese)

Keeling, C. D., T. P. Whorf. 1990. "TRENDS '90", A Compendium of Data on Global Change. Carbon Dioxide Information Analysis Center, Oak Ridge National Laboratory, Tennessee, USA. Cited in Organization for Economic Cooperation and Development (OECD) Environment Committee. 1992. Japanese edition of *The State of the Environment.* p. 14, Chuo Hoki Pub., Tokyo, Japan (in Japanese)

Keeling, C. D., T. P. Whorf, M. Wahlen, J. van der Plicht. 1995. *Nature,* **375**, 666

Lanly, J. P. 1982. *Tropical Forest Resources.* p. 106. Food and Agriculture Organization of the United Nations (FAO), Rome. Cited in Koizumi, H. 1990. *Kagaku Kogaku,* **54**, 22

Oikawa, T. 1990. *Kagaku to Kougyou (Chemistry and the Chemical Industry)* **43**, 1841 (in Japanese)

Marland, G. 1990. In Boden, T. A., P. Kanciruk, M. P. Farrel (eds.). TRENDS'90. A Compendium of Data on Global Change. Carbon Dioxide Information Analysis Center, Oak Ridge National Laboratory, Oak Ridge, Tennessee, USA, 37831-6335

Ministry of International Trade and Industry (MITI), Japan. 1990. *Report from Ministry of International Trade and Industry on Action Plan for Mitigating Global Warming*

Naftel, A, E. Moor, H. Oeshger, M. Staffer. 1985. *Nature,* **315**, 45. Cited in Tomisaka *et al.* 1990. *Kagaku Kogaku* **54**, 8 (in Japanese)

New Energy and Industrial Technology Development Organization (NEDO). 1992. Data in Information Center of NEDO, Japan

Organization for Economic Cooperation and Development (OECD). 1992. *The State of the Environment* (citations from the Japanese edition, 1992, p. 11, 14, Chuo Hoki Pub., Tokyo, Japan

Oikawa, T. 1989. *Modern Chemistry* **11**, 61 (in Japanese)

Okamoto, K. 1989. In Watanuki, K. (ed.). How to Save the Earth. p. 37. Kyoritsu Pub., Tokyo, Japan (in Japanese)

Ramanathan, V., R. J. Cicerone, H. B. Singh, J. T. Kiehl. 1985. Journal of Geophysical Research **90**, 5547. Citation from Komiyama, H. *et al.*, 1990. A Handbook of the Global Warming Issue, p. 46, IPC Publishers, Tokyo, Japan (in Japanese)

Redfield, A. C., B. H. Ketchum, F. A. Richards. 1963. In Hill, M.N. (ed.). The Sea—
   Ideas and Observations on Progress in the Study of the Sea, Vol. 2. p. 39. John Wiley
   & Sons, New York, USA
Riebesell, U., D. A. Wolf-Gladrow, B. Smetacek. 1993. *Nature*, **361**, 249
Scientific Committee on Problems of the Environment (SCOPE). 1986. SCOPE Report 29.
   How is Man Changing the Composition of the Atmosphere? Citation from Kaya, Y.
   1991. *A Handbook of the Global Environmental Engineering*. p. 493. Ohm Pub. Inc.,
   Tokyo, Japan (in Japanese)
Sekiya, A. 1990. in "A Handbook of the Global Warming Issue", in Japanese, ed. by
   Komiyama, H. *et al.*, p. 111, IPC, Tokyo
Tanaka, M. 1991. In Organizing Committee of 5th Symposium on University and Science
   (ed.). *Science of Global Environmental Change*. p. 172, Kubapuro Pub., Tokyo, Japan
Yoda, K. 1990. Data presented at a private seminar
Yoshida, A., M. Murase. 1990. Citation from Komiyama, H. *et al.*, 1990. A Handbook of
   the Global Warming Issue, p. 149, IPC, Tokyo, Japan (in Japanese)

3 THE WORLD'S ENERGY PROBLEM AND CARBON DIOXIDE EMISSIONS—THE CONVERSION OF PRIMARY ENERGY

In the previous chapter, the sources of carbon dioxide emissions were roughly classified into general categories. This chapter will be limited to a discussion of primary energy and the emissions due to the consumption of fossil fuels.

Primary energy is the most fundamental form of energy. The types of primary energy that will be considered in this chapter are those which generate considerable amounts of carbon dioxide, and also those that either produce no carbon dioxide, or else lead to the formation of only limited amounts; in addition, there will be an examination of the feasibility of converting to other forms of energy which produce less carbon dioxide.

[Secondary energy is energy which has been processed so as to make it easy to transport and use. The next chapter will examine the conversion of primary to secondary forms of energy, as well as improvements in its efficiency; it will also consider the various types of secondary energy, improvements in the efficiency of its transportation and consumption, and its conservation.]

There are also many forms of so-called alternative energy; the majority of these are primary forms of energy and thus will be treated in this chapter, but since hydrogen energy is a form of secondary energy, a discussion of this will be deferred until the next chapter.

3.1 SOURCES OF CARBON DIOXIDE EMISSIONS

Changes which Marland documented in 1984 are shown in Fig. 3.1; because of the oil shock, the consumption of fossil fuels in general (and oil in particular) was greatly reduced, and emissions of carbon dioxide from fossil fuels showed a marked decline.

The changes in carbon dioxide emissions for various regions of the world are shown in Fig. 3.2. The change in emissions in China (included in the data for Asian communist countries) was not continuous, due to unique internal political considerations, but in general there was a noticeable increase in emissions in developing countries and in the communist bloc (including the former Soviet Union and East Europe), but a decrease in OECD countries (i.e. the free world of advanced nations).

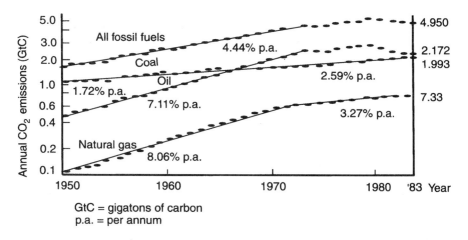

GtC = gigatons of carbon
p.a. = per annum

FIGURE 3.1 Changes in CO_2 emissions due to fossil fuel consumption [Marland and Rotty, 1984].

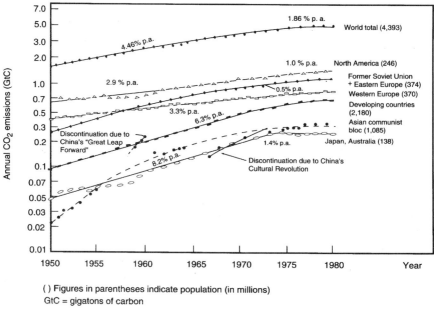

( ) Figures in parentheses indicate population (in millions)
GtC = gigatons of carbon
p.a. = per annum

FIGURE 3.2 Regional changes in CO_2 emissions from fossil fuels [Rotty *et al.*, 1984].

The differences between the various groups of countries are analyzed in Fig. 3.3, which also includes much more recent data (including the later increase in energy use after the fall in the price of crude oil). Taking the 1971 level of carbon emissions as 100, Fig. 3.3 shows subsequent changes with respect to per capita economic growth and to population; there seems to be a strong correlation

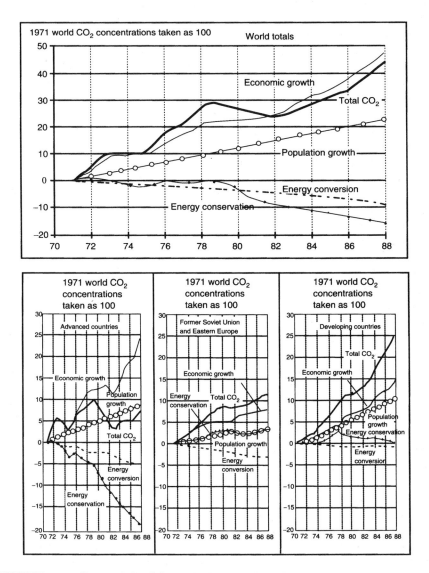

FIGURE 3.3 Changes in CO_2 emissions and in the relative contributions of factors determining CO_2 emissions [Ogawa, 1992].

between carbon dioxide emissions and economic growth. Furthermore, changes are seen both due to energy conversion (i.e. a change in the amount of carbon dioxide produced per unit of energy, or 'carbon intensity') and to energy conservation (i.e. the amount of energy needed per unit of economic activity, or 'energy intensity'). Energy conversion refers to a change from one type of energy to another. A change from thermoelectric power to nuclear power, or certain changes in the type of thermoelectric power employed (e.g. from oil-fired power stations to ones using natural gas), would lead to a reduction in the amount of carbon dioxide produced; conversely, conversion from oil to coal would increase the quantity of carbon dioxide formed. Energy conservation includes not only increased efficiency of energy use but also structural changes (such as a move away from energy-intensive industries); this might, for example, involve a change from manufacturing to knowledge-intensive industries.

According to an analysis by Ogawa (1993) global carbon dioxide emissions increased by 44% between 1971 and 1988, of which 8% occurred in advanced nations, 12% in the former Soviet Union and Eastern Europe, and 24% in developing countries. The drift of Ogawa's argument based on these figures is that although advanced nations had previously been the main source of carbon dioxide emissions, the recent increase indicates that the developing world is also playing a significant role in carbon dioxide production.

Let us now examine the matter in rather more detail. First, in the case of advanced nations, the increase in carbon dioxide emissions has been small relative to the considerable economic growth; that is, energy conservation and energy conversion have essentially covered the rise in economic growth. However, as was mentioned previously, the brake has since been taken off energy conservation and recently it has been possible to discern another rise in carbon dioxide emissions. In the case of the former communist bloc, one of the sources of the increase in carbon dioxide emissions has been the increase in the energy required per unit of economic activity. On the other hand, in developing countries, virtually no contribution can be observed from energy conservation and energy conversion; all of the increase in carbon dioxide has been due to population growth and economic growth. In conclusion, the worldwide totals for these categories show almost complete agreement between economic growth and the increase in carbon dioxide concentrations. There is also an argument which holds that increased carbon dioxide emissions are an inevitable consequence of economic growth, but on the basis of just Fig. 3.3 it appears that some caution must be exercised about accepting this assertion too strictly.

The transfer of energy-intensive industries from advanced nations to developing countries is one cause of the acceleration of energy conservation in advanced nations and the worsening of energy efficiency in developing countries. However, notwithstanding this fact, the events in the advanced nations have shown that even if the technological ability and investment capability is present, it is a sudden rise in energy costs that can be the great spur towards energy conservation.

3.2   REGIONAL VARIATIONS IN PER CAPITA CO$_2$ EMISSIONS FROM FOSSIL FUELS

As previously mentioned, advanced nations emit far more carbon dioxide than developing countries. For the various regions of the world, Fig. 3.4 compares the amounts of carbon dioxide released as a function of per capita energy consumption in 1987 with the average annual per capita emissions over the period 1800–1987 (i.e. the value obtained by dividing the cumulative amount released between 1800 and 1987 by the total population in each year). This shows that even among the advanced nations, North America has been by far the greatest emitter of carbon dioxide, both now and in the past. The consumption in the developing world has clearly been much less than in advanced regions, despite the marked increase in recent years. On the basis of Fig. 3.4, and notwithstanding the previously mentioned assertions of Ogawa, it is possible to acquiesce with the contention by developing countries that the problem of carbon dioxide emissions originates with (and should be addressed by) the advanced nations.

That being the case, to what precise extent do developing countries have the right to emit carbon dioxide? Let us consider the calculations by Fujii (1993), who made three assumptions:

(1) For three possible scenarios for the year 2100 (i.e. that the increased concentrations of carbon dioxide would be respectively restricted to 140 parts per million, 280 ppm, and 420 ppm), Fujii calculated the total emissions of carbon dioxide which would be permitted worldwide, and assumed that they would be distributed evenly between countries based on present population levels.

(2) Fujii also assumed that countries (or regions) whose cumulative production of carbon dioxide to date has been excessive would be considered as having had an advance from those permits.

(3) Fujii used the World Bank's model, which postulated that the world's population in 2100 will have stabilized at 10 billion. (It should be mentioned that this particular model is said to be on the relatively optimistic side).

Figure 3.5 shows the values obtained by dividing the permitted emissions for each country (or region) by the projected population totals over the period 1988–2100 (i.e. the annual per capita emissions). If countries are only permitted to emit carbon dioxide in accordance with the 140 ppm limit, this would in fact mean that North America would not be allowed to emit any carbon dioxide at all. On the other hand, Japan, Eastern Europe and Western Europe would be allowed to emit a total not so different from that for developing countries because in comparison with those nations they are not expected to experience so much population growth (and indeed may even register a decline in population).

It is necessary to also address the question of whether such a solution appears fair. For instance, the calculation produces results which are clearly too harsh for

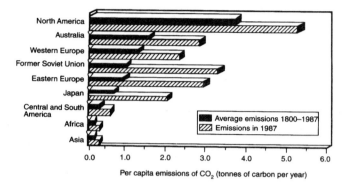

FIGURE 3.4 Per capita CO_2 emissions by region (for 1987, and annual averages over the period 1800−1987) [Fujii, 1993].

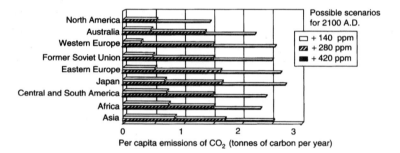

FIGURE 3.5 Future permissible per capita emissions of carbon dioxide (annual averages for 1988–2100) [Fujii, 1993].

North America. However, the most striking fact is that for developing countries the question of overpopulation is an extremely serious one.

3.3 ENERGY CONSUMPTION AND CARBON DIOXIDE EMISSIONS IN JAPAN

Since Japan is one of the leading countries in the debate about carbon dioxide and has the potential to make a significant contribution to a mitigation of the problem, and as Western readers may have few opportunities to appreciate a Japanese perspective, let us now consider Japan's position with respect to the carbon dioxide problem. The major difference from the pollution problem in the past is that approximately 50% of the emissions are due to private consumption

(Moriguchi *et al.*, 1993). That is, consumers themselves are the emitters—a point which also applies to other countries. This conclusion is reinforced by the data in Table 3.1, which provides a classification of emission sources. Although slight differences in the results are observed depending on the method of calculation, it is clear that industry is by no means the only source.

The changes in energy and carbon intensities (which are among the factors affecting the level of carbon dioxide emissions) in Fig. 3.6 for various advanced countries are classified by nation and by sector of the economy. In Japan, industry showed an annual 3% improvement in energy conservation between 1973 and 1979, a figure which rose to almost 6% in subsequent years. Energy conversion also became a major factor in the field of electricity generation, which was

Table 3.1 Sources of CO_2 emissions in Japan (Yoshida and Murase, 1990).

Sector of economy	Estimate by the Environment Agency	Estimate by The Institute of Energy Economics
Industry	33%	40%
Electricity and gas	27% (Electricity only)	31%
Transportation	23%	17%
Residential and commercial sector	17%	11%

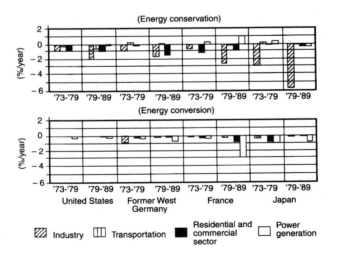

FIGURE 3.6 Reductions in CO_2 emissions in various economic sectors in Japan and in other advanced countries [Institute of Energy Economics (Japan), 1993].

primarily due to the conversion to nuclear power; in France this conversion was even more marked.

A comparison of carbon dioxide emissions for various countries by type of power generation (Fig. 3.7) reveals that the major sources of electricity generation in France and Canada are respectively nuclear power and hydroelectric power, and that in both cases the amount of carbon dioxide released per unit of electricity generated is small. With respect to thermal power generation, Japan produces the least amount of carbon dioxide of any country in the world per unit of electricity generated. Although one of the reasons for this is the difference in the amount of carbon dioxide produced per unit of heat generated by the different fossil fuels (a point which will be discussed later), the primary reason is the difference in efficiency of thermal power generation (Table 3.2).* Figure 3.8 shows the amounts of nitrogen oxides (NO_x) and sulfur oxides (SO_x) produced in various countries per unit of electricity generated by thermal power plants. Although this also is dependent on the particular type of fuel used, it is evident that Japan's environmental policies are advanced. Even excluding the consumption of electricity within the power plant itself for the purposes of desulfurization and denitration, Japan is still the most efficient country. It is clear that Japan possesses an extremely advanced technological capacity, both in terms of efficiency and environmental impact, which provides justification for asserting that Japan can make a major technological contribution to not only the carbon dioxide problem but also the world's environmental problems in general.

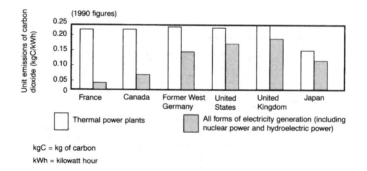

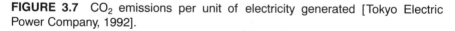

FIGURE 3.7 CO_2 emissions per unit of electricity generated [Tokyo Electric Power Company, 1992].

*A difference in efficiency of electricity generation of 1% translates as a difference of approximately 3% in the amount of carbon dioxide released. This is because the thermal efficiency of power generation itself is about 30–40% (Table 3.2); i.e. when the initial energy (taken as being 100%) is converted into electricity, only 30–40% of the energy is actually transformed into electrical form; if the efficiency is improved by 1% (e.g. from 33% to 34%) fuel consumption and carbon dioxide emissions are reduced by 1/33, which is equal to 3%.

Table 3.2 Comparison between nations of net thermal efficiency of thermal power generation, and rate of loss during distribution (Tokyo Electric Power Company, 1992).

	(Net) thermal efficiency at thermal power plants, % (1985)	Rate of loss during distribution, % (1989)
United States	32.7	6.3
Canada	32.0	9.1
United Kingdom	32.6	8.0
France	33.0	7.5
Former West Germany	33.2	4.0
Italy	35.1	7.6
Japan	36.4	5.6

Note: Net thermal efficiency is the gross thermal efficiency minus the loss in efficiency, resulting from power used within the plant itself.

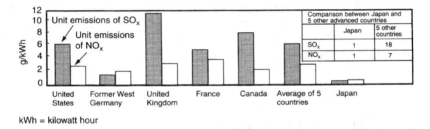

kWh = kilowatt hour

FIGURE 3.8 Emissions of NO_x and SO_x per unit of electricity generated at thermal power plants for various countries (1988 figures, with the exception of Canada, where the figures are for 1985) [Tokyo Power Company, 1992].

3.4 THE MAJOR SOURCES OF PRIMARY ENERGY

As stated earlier, primary energy is the most fundamental form of energy. Besides fossil fuels, primary energy includes nuclear power (both nuclear fusion and nuclear fission), solar energy (heat and light), and other forms of energy which have their origin in the sun's energy such as hydraulic power, wind power, energy produced by utilizing the temperature gradients within the ocean ("ocean thermal energy"), energy produced by utilizing the difference between densities ("concentration difference energy"), wave power, and biomass. In addition, there are other forms of energy within the earth itself (i.e. geothermal energy), and energy due to the orbital movement of the moon (i.e. tidal energy).

Although fossil fuels originate from biomass, which was originally produced by the sun and which then accumulated within the earth, it is not possible to regenerate this resource within the time scale available to human beings, and therefore this will be considered separately. With the exception of fossil fuels, all these forms of energy do not in principle lead to the formation of carbon dioxide. It should perhaps be noted, though, that since energy is used for the construction of nuclear power plants and in the treatment of nuclear waste, some people argue that nuclear power should be classified as a form of power generation that produces carbon dioxide, but in actual fact the amount involved is extremely small.

Energy other than fossil fuels and nuclear power is called natural energy or renewable energy. Starting with solar power generation, all forms of natural energy require the expenditure of energy for construction of the facilities needed for power generation, just as is the case with nuclear power; the problem is that the production of this energy necessitates the use of fossil fuels. However, adjustments can be made using the concept of energy payback time. The energy required for the construction of such facilities is analogous to a financial loan, whereby the loan is repaid over a period of years, although in this case there is no interest to be paid. Many theoretical models have been constructed, but in practice hydroelectric power and nuclear power have involved a much shorter payback time with respect to the lifetime of the facilities. Photovoltaic cells have a lifetime of 20–30 years and a current payback time of approximately 3–10 years, which means that the "loan" is still rather large. Nevertheless, from an energy point of view their introduction would be beneficial.

3.5 THE CHEMISTRY OF COMBUSTION

Energy is released when a material undergoes combustion. Fossil fuels are combusted in order to acquire this energy, even though the process involves the production of carbon dioxide.

$$C + O_2 = CO_2 + \text{energy (393.5 kJ/mol)} \tag{1}$$

$$2H_2 + O_2 = 2H_2O \text{ (water vapor)} + \text{energy (483.6 kJ/mol)} \tag{2}$$

$$2H_2 + O_2 = 2H_2O \text{ (liquid water)} + \text{energy (571.6 kJ/mol)} \tag{3}$$

[*Note*: 1 kilocalorie (kcal) $\approx$ 4.2 kilojoules (kJ).]

When hydrogen is combusted, energy is released in the form of heat with the formation of either water vapor [Eq. (2)] or liquid water [Eq. (3)]. Since it is possible to recover the heat of condensation that accompanies the chemical transformation described by Eq. (3), the amount of heat is greater (it is thus said to have a higher heating value).

All fossil fuels contain a mixture of carbon and hydrogen; other elements such as sulfur and nitrogen also undergo combustion with the release of heat, but the amount released is small. If fossil fuels existed which contained hydrogen but no carbon, it would be possible to burn them without the production of any carbon dioxide, and there would thus be absolutely no problem whatsoever. Unfortunately, however, such fossil fuels simply do not exist. As long as fossil fuels are burned, carbon dioxide will be given off.

3.6 THE CARBON DIOXIDE PROBLEM IS ESSENTIALLY AN ENERGY PROBLEM

Figure 3.9 reveals a considerable difference between countries in the use of primary energy sources. Certain countries (such as China) still rely heavily on coal. Although not specifically identified in Fig. 3.9, certain other countries (e.g. France) are very dependent on nuclear power. However, the total world production of nuclear power (plus hydroelectric power) amounts to merely a little over 10% of total energy production. The remainder (nearly 90%) is accounted for by fossil fuels; as long as these fossil fuels are combusted, carbon dioxide will be released at the same time as the fossil fuel resources are being depleted. The reduction of carbon dioxide emissions thus requires fossil fuels to be replaced by other sources of energy; in other words, the carbon dioxide problem can be viewed as being essentially an energy problem.

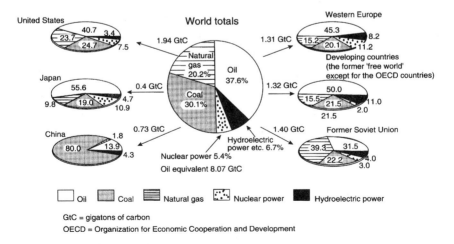

FIGURE 3.9 World consumption of primary energy, 1988 [Matsui, 1991].

3.7 THE AMOUNT AND CARBON CONTENT OF ENERGY RESOURCES

For various fossil fuels, Table 3.3 shows the hydrogen:carbon (H/C) ratio, the higher heating value, the amount of carbon dioxide released, and the size of existing reserves plus their oil equivalents. With coal in particular there is a range of types from carbon-rich to carbon-poor, and thus the figures shown in the table can only be deemed typical values. Natural gas has the greatest hydrogen content, followed by oil, whereas coal contains the maximum amount of carbon. The heating value per unit weight decreases in the same order, and the amount of carbon dioxide released (given in the table in terms of carbon equivalent) also increases in that order. Since coal includes hydrogen as well as oxygen and ash, the heating value of coal is less than that of carbon. In producing the same amount of energy, natural gas would release only half the amount of carbon dioxide that would be produced from coal.

Let us now consider the situation in which the use of coal and oil is completely replaced by the adoption of natural gas, which would be expected to bring about a considerable reduction in carbon dioxide emissions. Unfortunately, if the proven recoverable reserves are divided by annual production, we find that the original quantity of reserves would only last for 60 years; if in addition coal and oil were to be replaced by natural gas, supplies would only last for 15 years. Even the adoption of the figure for estimated ultimate reserves only leads to supplies being available for 27 years. Although recently there has been talk of natural gas supplies trapped in deep layers underground (which would mean that reserves are greater than had previously been believed), this does not affect the fundamental situation.

Next let us look at oil, which produces less carbon dioxide than coal. Unfortunately, supplies are once again limited, and would run out after 37 years. If coal were no longer used, even if we used all estimated reserves of both natural gas and oil (including uneconomic ones), supplies would be exhausted in 64 years. And that is working on the assumption that energy use does not increase over current levels, which simply does not seem realistic.

Coal is the most reliable fossil fuel. If all the coal could have been utilized, the energy problem would not have arisen for a thousand years; even with a slight increase in the amount used, supplies would have lasted for several hundred years.

3.8 THE PROBLEM OF CONVERSION TO LOW-CARBON FUELS

Let us consider what would happen if in the next half-century all or most of the primary energy sources were replaced by natural energy or nuclear power.

Table 3.3 Comparison of various fossil fuels with respect to CO_2 emission characteristics and amount of reserves (adapted from Inaba, 1991).

Characteristics and reserves	Carbon	Coal	Oil	Natural gas	Hydrogen	Total (oil) equivalent
*Higher heating value (kcal/kg)	7,800	7,000	10,000	13,000	34,000	—
H/C ratio (by number of atoms)	—	0.9	1.8	3.9	—	—
H/C ratio (by weight)	—	0.08	0.15	0.33	—	—
Amount of CO_2 emitted (gC/kcal)	0.13	0.11	0.078	0.058	0	—
Estimated ultimate reserves (E) (10^{12} t)	—	9.9	0.27	0.15	—	7.4
Proven recoverable reserves (R) (10^{12} t)	—	0.73	0.12	0.08	—	0.74
Annual production (P) (GtC)	—	3.5	3.0	1.4	—	$T=7.3$
Remaining years of supply (R/P, y)	—	200	40	60	—	100
Remaining years of supply (E/P, y)	—	2,800	90	100	—	1,000
Remaining years of supply (E/T, y)	—	950	37	27	—	—

T = Total annual fossil fuel production, oil equivalent.
*"Higher heating value" refers to when the combustion of hydrogen produces liquid water, rather than steam.

Considering the fact that in the last half-century Japan has switched her main source of electricity supplies from hydroelectric power to thermoelectric power, and also taking public opinion about nuclear power into consideration, the situation appears to be particularly difficult.

If it proved extremely problematical to convert primary energy sources into either natural energy or nuclear power, coal would have to be used. Which, then, should be used first—coal, or oil and natural gas? In order to simplify the discussion, let us consider the situation up to one hundred years into the future, with coal and natural gas each used for fifty years. What would the carbon dioxide concentrations be in one hundred years' time? Let us examine the situation if coal were used first, and then consider what would happen if natural gas were employed first.

At present, it will be hypothesized that only the same amount of fuel will be used, no matter whether coal or natural gas is actually utilized first. Consequently, whichever is used first, the total amount of carbon dioxide which is released would be the same. As shown in Fig. 3.10(a), if it is assumed that the oceans have no effect, then the prior use of natural gas would not greatly increase carbon dioxide concentrations; the subsequent use of coal would then lead to a rapid rise in carbon dioxide levels. On the other hand, the prior use of coal would cause a rapid increase in carbon dioxide concentrations, which would then stabilize with the conversion to natural gas. However, the final amount would be the same in both scenarios.

The above assumption that the oceans do not play a role is of course a false one; the oceans absorb carbon dioxide because of the increased concentrations of carbon dioxide in the atmosphere. If coal were used before natural gas, the atmosphere would have to retain higher concentrations of carbon dioxide for a long period of time. Accordingly, the oceans would presumably absorb a considerable amount of

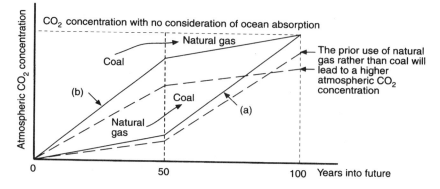

FIGURE 3.10   Coal or natural gas: which should be used first?
(a) With no consideration of ocean absorption
(b) With consideration of the absorptive capacity of the oceans, which increases with increasing CO_2 concentrations.

this carbon dioxide. Incorporating the effect of the oceans into the discussion, the result would be that if coal were used first, the final carbon dioxide concentration would be less than if natural gas were used first [Fig. 3.10(b)], although it would of course mean that for most of the hundred years considered the carbon dioxide concentration would be high. This argument considers only the effects of the oceans, but it must be remembered that in reality, as was described in the previous chapter, land vegetation would also have a similar effect to that of the oceans.

It needs to be realized that the conversion to low-carbon fuels is a rash policy for humans to adopt. Even compared to other fossil fuels, coal is difficult to handle, but countries with the technological capacity and economic power should use coal in an efficient manner that does not lead to the production of pollution. Developing countries are currently using coal with comparatively little market value in an inefficient and highly-polluting manner; this low-grade coal is difficult to use, and often has a high sulfur and nitrogen content. This is causing an environmental problem on a broad scale which is not limited to just the questions of carbon dioxide emissions but also includes the oxides of sulfur and nitrogen (SO_X and NO_X) which lead to acid rain, and the emissions of N_2O, which is contributing to the destruction of the ozone layer. Rather than the financially powerful advanced nations adopting such simple policies for reducing carbon dioxide emissions, it would be better for these resources to be directed toward the countries which are developing technologically, with the advanced nations devoting their efforts to developing techniques to combat the problems posed by carbon dioxide. It is also necessary for advanced nations to quickly establish alternative ways of handling renewable energy (including conversion and transportation) while minimizing the use of fossil fuels.

One proposal for restraining the use of fuels that produce large quantities of carbon dioxide is the introduction of carbon taxes. A 'carbon tax' is a tax which is levied in accordance with the consumption of carbon. If the amount of carbon used is small, the tax is low. For the same amount of energy produced, the tax levied on natural gas would be half that levied on coal. This author is certainly not opposed to a carbon tax in itself, but from the above argument it would seem that the tax should really be a fossil fuel tax. Indeed it might be better to pursue a policy such as imposing an energy tax and then recycling this as a tax rebate and as a subsidy for the development of renewable energy.

It is meaningless to merely instigate a competition between the various forms of fossil fuels. A round peg should be placed in a round hole.

3.9 THE HYDROCARB PROCESS

Let us consider whether it is possible to obtain energy from the abundant energy resource of coal without the release of carbon dioxide. One proposal for realizing this is the Hydrocarb process (Steinberg, 1989), in which only the hydrogen in the coal would be extracted, with the remaining carbon being reburied.

Let us make a rough estimate using the figures in Table 3.3. If it is assumed that the ash content is zero and that all the hydrogen is extracted, then the amount of hydrogen contained in 1 kg of coal is equal to $0.08/(1+0.08) = 0.074$ kg; that is, it amounts to a mere 7.4%. Multiplication of this by the heating value of hydrogen means that 1 kg of coal would yield 2,500 kcal of heat. In practice, however, ash and oxygen are also present in coal. Not only does the oxygen not lead to the production of heat, but since the oxygen is also already weakly combined with the hydrogen in a water-like state, part of the heating value of the hydrogen included in the above calculations is in practice unrealizable. For such reasons, only 23% of the chemical energy in coal can theoretically be converted into heat; however, when the heat losses that occur in the process are also incorporated, the figure is believed to fall to below 20%. In this case, the proven recoverable reserves would fall to 40 years' supply (based on the division of reserves by annual production, i.e. R/P); a calculation based on the ratio of estimated ultimate reserves to total reserves (E/T) would show that if all energy were to be derived from coal, then the total estimated reserves would last less than 200 years.

This is therefore a technology that cannot be employed until such time as absolutely no carbon dioxide can be released and only coal can be used. If the remaining carbon could be used as carbon products, this would pose no economic problems; however, even if it might appear that carbon could be used as a raw material, in many cases it would actually end up in the form of carbon dioxide. Also, the carbonaceous raw material often serves as an energy source. For example, the cokes used in the steel industry are themselves converted in the blast furnace into carbon dioxide with the production of energy. This is also then included in the calculations as primary energy. The only applications that would probably not be accompanied by the production of carbon dioxide would be restricted to the manufacture of certain carbon fibers etc., and even then the total amount involved would be by no means large. Even if the carbon were to be used in such a way, there would be no change in the situation concerning the decrease of energy resources.

It has recently been suggested that a combination of the Hydrocarb process with the use of biomass would apparently provide a considerable amount of energy without the production of carbon dioxide. However, this is basically nothing more than the combined use of coal and biomass; it produces absolutely no benefit, and merely amounts to playing around with the figures.

A further alternative proposal suggests a combustion method that produces soot. However, the fundamental thinking is the same as that which underlies the Hydrocarb process, and the proposal therefore offers no particular attractions.

A technique which is intermediate between the Hydrocarb process and the use of every part of the coal is that of carbonization. The products of carbonization are hydrogen, a gas which has properties similar to natural gas, and a tar with a composition similar to oil, leaving behind carbon- and ash-derived cokes (as mentioned before) or so-called 'char'. In this case, approximately one-third of the energy of the coal can be used, but unfortunately the energy source is also depleted to about one-third.

In each of these cases, assuming that only a fraction of the value will be extracted, there would be an increase in energy costs, and this is therefore likely to lead to the development of a new energy substitute.

3.10 BIOMASS

The term biomass refers to a living body or a product of a living body; a typical example is firewood. This section will consider the use of biomass as a form of energy.

To be precise, biomass is produced from water and atmospheric carbon dioxide by the utilization of solar energy. It should be noted that doubt surrounds the question of whether there is a net release of carbon dioxide due to the combustion of biomass. The gist of the argument behind this doubt is that although carbon dioxide is of course released, in effect the net total emission may be zero since the released carbon dioxide was originally present in the atmosphere before it became fixed by the biomass. There is, though, a weak point in this argument. That is, if a tree is felled, and then the land where the tree used to be is left as it is, it will in effect mean that carbon dioxide is being released. In order to reduce carbon dioxide emissions to zero, it is necessary to replant a tree after one is felled, and thus return the land to the original state.

Let us now compare the use of biomass with the use of fossil fuels. Even fossil fuels are materials that were formed by the fixation of atmospheric carbon dioxide. If biomass is referred to as a form of renewable energy, then the same should seemingly apply to fossil fuels. In the final analysis, however, biomass can only be regarded as renewable energy (i.e. energy that can be regenerated) if the term is limited to those products which are grown and harvested by humans themselves, and to the use of surplus products resulting from the actions of stable ecosystems.

The great difference between biomass and fossil fuels is that it is impossible to synthesize coal or oil within a few decades, whereas the synthesis of biomass is simple from the technological point of view. Carbon can be fixed naturally by the cultivation of saplings alone, which can be achieved by merely providing an environment for trees to grow.

When discussing the use of biomass, it is necessary to distinguish between actions which destroy ecosystems while releasing carbon dioxide; actions which

Table 3.4 The use of firewood as a source of energy in 1985 (United Nations, 1986).

Amount used and rate of increase	Firewood	Crude oil	Natural gas	Coal	Lignite
Amount used (gigatons of coal equivalent)	0.564	3.792	2.052	2.505	0.442
Rate of increase (%)*	2.9	0.0	3.0	2.9	3.3

*Average rate of increase over the period 1975–85, relative to total use in 1975.

under regular supervision constitute one form of use of solar energy; and the fixation of carbon that results from the nurturing of forests. If carbon taxes were to be instituted, these three situations would be treated differently, i.e. a carbon tax would be imposed in case #1, no tax would be levied in case #2, and a tax rebate would be given in case #3.

Special mention should be made of firewood, since this accounts for almost 10% of total energy use. Using statistics compiled by the United Nations, Table 3.4 compares the amount of firewood used as an energy source with the amounts of other fossil fuel resources. With the exception of oil, the use of which declined following the oil shock, the use of firewood has increased by approximately the same rate as other fossil fuels. There would of course be no problem if this were included in the second category above, i.e. where the activity was regularly supervised; however, if one considers the decrease in forested land and the increase in desertification, there are serious misgivings about whether firewood deserves to be classified in this way—it might be more appropriate for firewood to be placed in category #1, i.e. an activity that causes destruction of an ecosystem.

Several countries have recently introduced carbon taxes, and these have become the subject of much debate (this will be discussed further in Chapter 8). In one of those nations, Sweden, the existence of a carbon tax is said to have led to the relocation of energy-intensive industries to other countries and the promotion of a transition from coal to biomass. However, the cost of domestically produced biomass is high, resulting in a reliance on imports. This somehow does not seem to be the most desirable state of affairs.

3.11 THE DEVELOPMENT OF RENEWABLE ENERGY SOURCES

In order to halt or reduce the use of fossil fuels without altering the total amount of energy consumed, it is necessary to utilize either energy which originates in the sun or earth (i.e. natural energy) or nuclear energy. The thermal energy of the sun has been utilized on both small and large scales; a small-scale example is the use of solar-heating devices for warming water, and a large-scale example is solar thermal generation of electricity (although the technology has only reached the pilot plant stage). On the other hand, the small-scale use of photovoltaic cells has already put into commercial operation, although it has only been realized in certain localized or specific instances. Large-scale projects for generating electricity from photovoltaic cells have not yet progressed beyond the pilot plant stage, with cost in each case proving to be a major problem. At the present time, it has been calculated that the electricity obtained from solar photovoltaic cells is either similar in cost or slightly more expensive than the electricity obtained from solar thermal radiation; however, it is anticipated that future efficiency improvements and cost reductions will lead to solar photovoltaic energy being preferred.

It is believed that economies of scale will contribute greatly to expanding the use of electricity generated from photovoltaic cells. At present, this form of electricity is several times more expensive than other commercialized primary energy sources. If the production of photovoltaic cells were to be increased by a factor of several dozen times, the cost would be cut in half. Furthermore, advances are being made in production technology and efficiency is expected to increase.

The efficiency of energy conversion in polycrystalline photovoltaic cells is high, but the cost of the materials is also high. Work is currently proceeding on proposals concerning a new concept for manufacturing polycrystalline silicon, and research is also being carried out on the method of producing ribbon crystals; in addition, researchers are attempting to verify the effectiveness of the various methods and to further improve them. In the case of amorphous silicon, the main goals are to increase both the efficiency of energy conversion and the durability of the cell. In addition, the Electricity Utility Industry Law in Japan was amended in January 1992 so that electricity generated from photovoltaic cells installed on the roofs of ordinary homes could be sold to electricity companies. This was an epoch-making event for Japan, but it is interesting to compare the situation in Japan with that in Germany. Germany may be regarded as a leader in the area of establishing the economic and political means for promoting the use of solar energy; however, even though the cost per house of the photovoltaic cell in both countries is the same at 4 million yen (about $40,000), the cost of an inverter* is far greater in Japan ($40,000 in Japan compared to $8,000 in Germany). The costs of installation and taxes further increase the total by $50,000 in Japan and $18,000 in Germany. Consequently, the total cost in Germany ($66,000) is only

---

*A photovoltaic cell produces a direct current of electricity, and it is necessary to convert this to an alternating current at 100 V. In order to use it both for household purposes and for precision machinery, it is generally necessary to have a constant frequency and voltage.

about half of that in Japan ($130,000). [The difference in cost is attributed to Japan requiring extra equipment on safety grounds.] On top of that, in 1993 in Germany a sum of $46,000 was offered as a subsidy, reducing the individual's investment to $20,000, which was a mere fraction of what it would be in Japan. Japan, which is a leading producer of photovoltaic cells, clearly needed to adopt a new policy.

In 1994, the situation in Japan was partially alleviated by the introduction of a subsidy amounting to 33% of the total cost; however, this is inadequate in view of the German position, and the fact that only a limited number of individuals qualify for the subsidy.

Besides solar energy and the previously mentioned biomass, other forms of natural energy which are under study include hydraulic power, wind power, wave power, tidal power, geothermal energy, energy due to temperature gradients (ocean thermal energy) and concentration gradients (concentration difference energy). Japan has achieved positive results with hydroelectric power and geothermal energy, but for the future there are problems regarding location. Potential sites exist in various parts of the world, but there are problems with transportation of the generated energy to the sites of consumption. Although various problems exist involving biomass (as discussed in the previous section) there are examples of large-scale energy production using biomass (e.g. in Brazil, where sugar cane has been converted into ethanol and used as a fuel for transportation). However, as will be discussed in Chapters 5 and 6, because the efficiency of converting solar energy to biomass is poor compared with the efficiency achieved when solar energy is converted into electricity by means of photovoltaic cells, and since further energy and financial costs are involved in the later treatment of biomass, the technique would only be feasible under certain specific conditions.

In advanced countries there are greater cost restrictions than in developing nations on the use of both timber and microorganisms. However, there are many aspects involved in the use of natural energy, including the use of surplus goods. If policies are adopted that include measures such as subsidies which are aimed at expanding usage, there should be an increase in the number of areas in which the techniques will be feasible, even when the initial investment is rather large. It is important to adopt the stance of "Use what can be used."

3.12 NUCLEAR POWER AND SAFETY CONSIDERATIONS

It would be nice to think that it is possible to prevent incidents that produce widespread environmental pollution, whether this is the result of an accident (such as the one at the Chernobyl nuclear power plant) or whether it is caused by war (e.g. the oil spills that occurred during the Gulf War). In order to achieve this, though, it would be necessary to adopt policies which are truly infallible. However, if we have to accept that we can never have a 100% guarantee, how should we approach this question? While this is perhaps a postscript to the earlier discussion, let us consider which is safer—nuclear power or fossil fuels.

Certainly any accident involving nuclear power would be irreversible. However, it should also be remembered that a number of similar or even greater accidents and incidents of environmental destruction have occurred with the use of fossil fuels such as oil. In the Gulf War there was spillage of large quantities of crude oil which caused major contamination of the ocean, and considerable amounts of sulfur dioxide and soot were also emitted; the large quantities of exhaust gases emitted from automobiles (and NO_x in particular) have caused a serious problem; and mining activities alone have also claimed many victims. Indeed, the very existence of fossil fuels may cause problems—it is even possible that the Gulf War might not have taken place if oil had not been present in the area.

It is thus extremely difficult to clarify which is safer. Similarly, it is difficult to reach a definite conclusion about which of nuclear power and fossil fuels is the cheaper method of generating electricity because here again, many arguments are involved. However, there are doubts about whether the cost and energy evaluations have adequately reflected that the role of nuclear power can only be that of a baseload* (i.e. the use of fossil fuels will also be necessary as a supplementary source of energy). However, with respect to cost, let us assume that, as is generally suggested, nuclear power is slightly cheaper than fossil fuels. Or at least, incorporating the above doubts, that the costs are approximately equal.

Let us postulate that statistical and probability studies show that nuclear power may be safer than fossil fuels. Nevertheless, public opinion is against nuclear power; let us now examine the reasons for this.

---

*The baseload can be considered to be the minimum required amount of electricity. In Japan electrical consumption reaches a peak during the hot summer days; on the other hand, the consumption of electricity is low at night. In other countries, there will also be variations resulting from local conditions. In essence, the amount of electricity generated needs to be varied in accordance with the amount consumed, but it is difficult to do this with nuclear power. However, once a power plant has been constructed the unit price of generated electricity is extremely low. Thus the amount corresponding to the minimum level of consumption (the baseload) could be provided by nuclear power, with the remaining fluctuating requirement being fulfilled by pumped storage power plants, that is, hydroelectric power plants which use the electricity generated at night to transport water from the bottom of a dam to the top, from where it is released whenever necessary to regenerate electricity. Hydroelectric power plants are the best for adapting to load variations, but the use of only hydroelectric power generation is insufficient, and consequently at certain times it is necessary to adjust power output by using power from coal-, oil-, and LNG-fired power plants. If nuclear power could easily deal with the load variations, there would not be such a great need for pumped storage power plants, and the costs of constructing thermal power stations would fall. (While mentioning this topic, it is interesting to note that the disaster at Chernobyl occurred during an experiment involving load variations.)

Consider the analogy of a lottery. If a ticket costs $10, people will still buy tickets even if the average return is only $4 or $5. That is because they are dreaming of winning a million dollars.

Although nuclear power is the exact reverse of the lottery situation, it still reflects the same way of thinking. Because of what happened at Chernobyl, it seems that we are reluctant to adopt nuclear power because we do not want to choose only on the basis of profit and loss. This stems from a belief that if something went wrong, the results would be catastrophic, despite the fact that such an event would be highly improbable. For these reasons, it appears that it would be extremely difficult to persuade public opinion to side with nuclear power out of hand. On the other hand, however, as is evidenced by the resumption of operations at Chernobyl, it is currently unrealistic to reject the use of nuclear power out of hand. In the end, the most likely scenario is that nuclear reactors which are currently in operation will continue to function, whereas the construction of any new reactors would face considerable opposition.

Present-day nuclear power plants are based on nuclear fission; in theory, however, nuclear power generation should also be possible using nuclear fusion. Nuclear fusion is a technology that has been dreamt of as a source of nuclear power, but it is not likely to be realized in the near future. Fast breeder reactors are also a long way off from becoming a reality. There are also inherent dangers with these technologies, and it goes without saying that it will be necessary to give full consideration to the safety aspects involved.

3.13 THE WARMING EFFECTS OF POWER PLANTS

Let us analyze the amounts of carbon dioxide which are emitted ("unit emissions") for each kilowatt hour of electricity produced from a range of primary energy sources. [One kilowatt hour (1 kWh) is the quantity of electricity consumed when one kilowatt of electricity is used for a period of one hour]. The unit emissions include not only emissions from the fuel itself but also the carbon dioxide released as a result of using electricity or fossil fuels for the construction of the plant facilities (in each case it is assumed that the facilities would last for 30 years); for the operation of the plant itself; and for transportation. Furthermore, in the case of coal and natural gas the figure includes an estimate for the methane that is released at the time of extraction. The natural decomposition of methane is also considered, with the warming potential over a period of 100 years being estimated at twenty times that of carbon dioxide (Fig. 3.11). In Fig. 3.12 the vertical axis represents the reciprocal of the amount of carbon dioxide released; that is, this figure is inversely proportional to the amount of carbon dioxide produced. The horizontal axis shows the energy balance of the

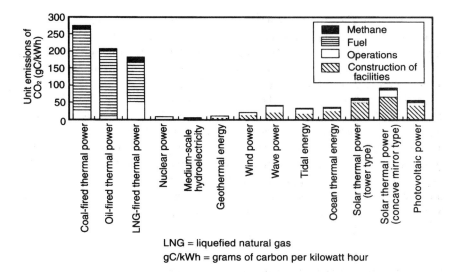

LNG = liquefied natural gas
gC/kWh = grams of carbon per kilowatt hour

FIGURE 3.11 Warming effect of electrical power plants [Uchiyama and Yamamoto, 1992].

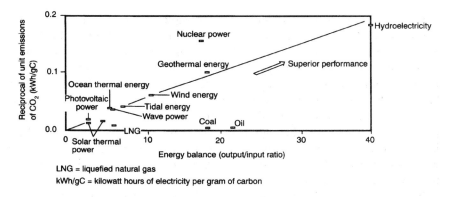

LNG = liquefied natural gas
kWh/gC = kilowatt hours of electricity per gram of carbon

FIGURE 3.12 Warming effect and energy balance at power plants for various energy sources [Uchiyama and Yamamoto, 1992].

process (i.e. the ratio of the amount of electricity generated to the amount of energy input, excluding the energy present within the fuel itself). For both the horizontal and vertical axes, the greater the value, the more advantageous the process.

Although the energy balance is favorable for all fossil fuels, they all emit large quantities of carbon dioxide, which is almost entirely the result of combustion (Fig. 3.11). The amount of carbon dioxide released decreases in the order of coal, oil and LNG (liquefied natural gas), but in the case of LNG although the emissions from the fuel itself are low there is not much difference from the value for oil if one includes the leakage of methane and the carbon dioxide which is formed as a result of using energy at the time of liquefaction. The reason for the rather poor energy balance of LNG relative to coal and oil is that it includes a large input of energy for liquefaction.

On the other hand, although there are no emissions from fuels based on fossil energy, all forms of natural energy involve considerable emissions resulting from the construction of facilities and transportation. With the exception of hydroelectric and geothermal power plants, the magnitude of emissions from the construction of facilities means that it is necessary to build large facilities. This necessitates not only an improvement in efficiency but also an extension of the lifetime of such facilities, and means that an increase in the rate of operation will have a considerable effect on the global warming problem.

3.14   ENERGY PRICES AND THE UTILIZATION OF UNUSED ENERGY

Let us now consider waste products such as garbage and also the use of previously unused energy. There are actual cases of the methane emitted from garbage landfills being used as a fuel. The heat which is given off in garbage incinerators which produce the highest temperature gases constitutes the easiest form of heat recovery. Thermal power plants also generate considerable unused heat, which forms a potential energy source that is believed comparable to the amount of heat needed for all the heating-cooling units in the Tokyo metropolitan area. In addition, although the temperatures are certainly not high, there is the heat released from water following sewage treatment, river water, sea water and subways; by the use of heat pumps, it is possible to utilize this heat for heating-cooling units.

In practice, even in the areas surrounding Tokyo there are several places where released thermal energy is used for regional heating and cooling, or where plans exist for this. These techniques should be developed further in the future, and the author hopes that policies will be devised to support such actions. However, the overriding concern is that energy be recognized as a valuable commodity; based on this, we need to look very seriously at how energy prices are determined.

References

Fujii, Y. 1993. Evaluation of countermeasures against carbon dioxide in energy systems (doctoral thesis). The University of Tokyo, Japan (in Japanese)

Inaba, A. 1991. Cited in Kaya, Y. 1991. *A Handbook of Global Environmental Engineering*, p. 497, Ohm Pub. Inc., Tokyo, Japan (in Japanese)

Institute of Energy Economics (Japan). 1993. *Energy and Resources* **14**, 12 (in Japanese)

Marland, G. and R. M. Rotty. 1984. *Tellus* **36B**, 232. Cited in Kaya, Y. 1991. *A Handbook of Global Environmental Engineering*, p. 494, Ohm Pub. Inc., Tokyo, Japan (in Japanese)

Matsui, K. 1991. *How to Read and Use Energy Data*. Denryoku Shinpo Pub., Tokyo, Japan (in Japanese)

Moriguchi, U., Y. Kondo and H. Shimizu. 1993. *Energy and Resources* **14**, 32 (in Japanese)

Ogawa, Y. 1992. *Petrotech* **15**, 817 (in Japanese)

Rotty, R.M., G. Marland and N. Treat. 1984. U.S. Department of Energy DOE/OR/21400-2. Cited in Tomisaka, Y. 1990. *Kagaku Kogaku* **54**, 8 (in Japanese)

Steinberg, M. 1989. *Proceedings of the International Conference on Coal Science, 1989*, p. 1059. International Energy Agency, 1059, and Brookhaven National Laboratory (New York) BNL Report, BNL 42228

Tokyo Electric Power Company. 1992. *Report on "Action Plan for Environment"* Tokyo, Japan

Uchiyama, Y. and H. Yamamoto. 1992. Greenhouse Effect Analysis of Power Generation Plants. Research Report of Central Research Institute of Electric Power Industry, Y91005, pp. 34–35

United Nations (Department of International Economic and Social Affairs, Statistical Office). 1986. *1985/86 Statistical Yearbook* (Vol. 35), United Nations, New York

Yoshida, G. and M. Murase. 1990. Cited in Komiyama, H. *et al.* (eds.). 1990. *A Handbook of the Global Warming Issue*, p. 149, IPC, Tokyo, Japan (in Japanese)

4 ENERGY CONSERVATION AND IMPROVEMENTS IN ENERGY EFFICIENCY: SECONDARY ENERGY SYSTEMS

Chapter 3 examined primary energy, the most fundamental form of energy; Chapter 4 will review secondary energy systems, which are those employed when primary energy is transported and delivered for use by the consumer.

4.1 ENERGY EFFICIENCY

As a result of twice suffering from 'oil shocks' Japan has attained a level of energy efficiency which she can be proud of; if this level of efficiency could be attained throughout the world, it might be possible to cut carbon dioxide emissions to less than half their current level. Energy efficiency can be defined in a number of ways, from net thermal efficiency* at the site of generation to the unit consumption rate of energy** for various manufactured products. This section will look at the questions of energy efficiency; first, though, let us begin by considering the different forms of secondary energy.

A table which shows both the forms in which energy is introduced into a country and the forms in which it is actually used is termed an energy balance table. Such a table contains a wealth of data; here, though, we will consider only the points that are important with respect to the production of secondary energy from primary energy.

An energy table for Japan would show that for electricity and gas, the original primary energy is converted into secondary energy by the energy utilities. In 1988, the total primary energy supply amounted to 4,320 trillion kilocalories; following conversion by either the electricity companies or by companies or individuals using their own generators, nearly 40% of this (1,660 trillion kilocalories) was used in the form of electricity. This is termed the electrification rate on a primary energy basis. Approximately 38% of the primary energy is

*During the generation of electricity, a certain amount of electricity that is generated within the plant is diverted and used as power within the plant itself for purposes such as desulfurization. The net thermal efficiency is equal to the gross thermal efficiency of electricity generation minus the loss due to such processes.

**The unit consumption rate of energy is the amount of energy needed to manufacture one unit of final product.

91

transformed into electricity (thermal efficiency), with the remainder being released as heat, which would be considered in the energy table as energy consumed by the electricity company. In contrast, the amount of primary energy converted into town gas is less than 10% of that used for generating electricity, but almost all of this energy is actually converted into town gas. Most of the remaining secondary energy is in the form of petroleum products obtained from crude oil (i.e. gasoline, kerosene, the heavy oils used in factories, and small quantities of chemicals). The majority of the secondary energy that is obtained from coal is coke used in the steel industry; a very small amount of heat is derived as secondary energy. Hydrogen and methanol are likely future forms of secondary energy.

During the Second World War, the energy used for transportation (such as gasoline) was more precious in Japan than electricity. Since the hydrogen content of coal is small in comparison with oil (Table 3.3), electricity was used to produce hydrogen; this was then added to coal, which in turn was liquefied in order to produce gasoline. This rather surprising procedure was actually implemented successfully by the Japanese in Manchuria.* Such equipment again saw the light of day at the time of the oil shock, although that was because of the fear that oil imports would come to a complete stop. Nevertheless, electricity is also produced from fossil fuels today. But even if coal liquefaction is being carried out on a temporary basis compromises are still needed from the point of view of energy.

Times are indeed changing. We are entering an era in which anything can be produced if electricity is available. For example, take the case of the so-called "Biosphere II" experiment, in which eight scientists (four men and four women) were confined within a closed space in the American desert. All materials were recycled. As long as energy was available, oxygen could be obtained from carbon dioxide (or, rather, as long as the cost posed no problem).

We have in fact now reached the point where we cannot live without electricity. The degree of electrification has increased with the passage of time—but how much energy is needed to generate electricity? It must be remembered that it is only possible to convert 38% of primary energy into electricity.

Let us take the example of heating–cooling units (air-conditioners which are also able to act as heaters). The heat pumps enable an amount of heat to be pumped in from the outside equal to more than double the quantity of electrical energy needed to operate the machine, even though electrical heaters only produce an amount of heat equivalent to the energy present in the original electricity. On the other hand, the hot water that it is used for baths certainly does not contain large amounts of "available" energy (here, "available" energy refers to energy that can be converted into work, i.e. mechanical energy). Accordingly,

---

*Manchuria is an area of north-eastern China in which Japan installed a puppet government before the start of the Second World War.

the discussion therefore must not be limited to only the quantities of energy involved—we must also consider the quality.

Despite the fact that the maximum efficiency of energy conversion obtained when electricity is produced by the combustion of fossil fuels is 40%, the rate of consumption of electricity has continued to increase, and is expected to grow still further in the future. That is because the energy in the form of electricity is more available for use than is at first apparent. The energy contained in fossil fuels is also an extremely available form of energy, even if not to the same degree as electricity.

The question of devising a more efficient fossil energy system is currently under investigation, with a variety of ideas being considered, for example, the use of energy cascades and energy cogeneration* (these are discussed in Section 7). In both cases, the aim is to use the energy repeatedly, or to portion out the energy in the form required at the time, but doing so with the minimum reduction in the availability of the energy. However, at such times it is impossible to discuss how efficiently the energy is used if only the conversion efficiency is considered. If the 60% of the energy that is dissipated in the form of heat during the production of electricity is used efficiently, and if the 40% that is converted into electricity is used for heating–cooling units, then the overall efficiency of energy use would exceed 100%. The use of cascades could also lead to a figure above 100%. Even if the energy is used once, the figure does not fall. That said, however, in practice the efficiency does actually decrease. The degree of availability varies according to where and for what purpose the secondary energy is used.

During transportation and use, the form of the energy is changed and part of the "available" energy is lost. The way in which this is controlled could be termed a "showplace" of human resourcefulness. If it is accepted that humans must rely on fossil fuels, then it follows that the amount of primary energy which is available to humans is limited. (That is true despite the fact that for hundreds of years there were no problems—a state of affairs that persisted up until the time that the problems related to carbon dioxide and global warming became apparent). If primary energy did not exist, there would also be no secondary energy.

As long as fossil fuels are used as primary energy, it will be impossible to reduce carbon dioxide emissions to zero. Even if the efficiency of conversion were doubled, the amount of carbon dioxide released would only be cut by 50%. On the other hand, no matter how low the conversion efficiency, the use of only solar energy would completely preclude the emission of carbon dioxide at the source.

---

*Cascades involve the repeated use of energy; the thermal energy which is utilized as electricity or as a means of attaining high temperatures is employed for various purposes at successively lower temperatures.

"Cogeneration" is a single system that produces both electricity and heat, and which is closely related to the practical use of unused energy. In order to increase the overall energy efficiency, the relative proportions of heat and electricity produced are regulated.

However, without a method of transportation it would be impossible to utilize solar energy. The reason we must use fossil fuels even though we are drenched with sunlight (which is the source of so much primary energy) is not just the fact that the conversion of solar energy into electricity requires the expenditure of energy and financial resources. An extra factor is that the energy produced is difficult to collect and transport.

4.2 INCREASING THE EFFICIENCY OF FOSSIL FUEL–ELECTRICITY CONVERSION

The generation of electricity is one area in which technological innovations are expected to increase the efficiency of converting fossil fuels into secondary energy. As stated earlier, in Japan the net thermal efficiency is currently of the order of 38% (the world's most efficient system of generating electricity), but the value varies greatly between different countries (Table 3.2). Figure 4.1(a) shows a steam turbine which is employed at a thermal power plant that uses pulverized coal*

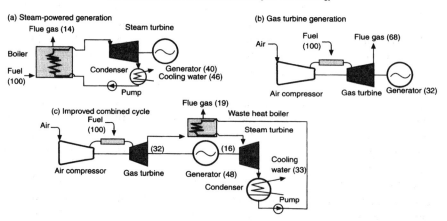

FIGURE 4.1 Differences between conventional steam-powered generation of electricity and electricity generated by combined cycles [Tokyo Electric Power Company 1992].

*In a system using pulverized coal, the coal is crushed into extremely small pieces and then combusted at a temperature of approximately 1,400°C in a stream of air; the heat is collected by water pipes located in the upper part of the facility, and this is subsequently converted into electricity by a steam turbine.

(a system which is employed at most coal-fired thermal power plants). Although the efficiency at a relatively new power plant using this system in Japan is approximately 38%, in China the efficiency falls to 28% even for the same thermal power (Table 4.1).

Is it possible to further increase the efficiency? In a steam turbine, thermal energy is converted into electricity at a temperature of approximately 500°C. The efficiency of the steam turbine itself rises with an increase in temperature. So-called "supercritical pressure steam power generation", in which operations are performed at temperatures and pressures even greater than normal, is at the stage of actualization; if the temperature and pressure are increased still further, this is termed "super supercritical pressure steam power generation". Later in the book we will consider the use of this technique in conjunction with pressurized fluidized bed combustion.

In steam turbines the energy is converted once into thermal energy, so in order to use the energy efficiently, it is necessary to have a technique that directly utilizes the kinetic energy of the high-temperature gas. One possible scheme is to directly introduce energy which is still in a highly efficient state (such as hot fossil fuel gases) into a gas turbine (Fig. 4.1(b)); after the maximum possible amount of electrical energy has been extracted, the remaining energy (in the form of thermal energy) is transferred to a boiler. Here the waste heat is recovered and used to make steam, which is then used in a steam turbine to generate more electricity. In other words, this is a combined cycle (Fig. 4.1(c)*). A plant employing natural gas as the raw material for a combined cycle is already in

Table 4.1 Generating efficiency and carbon dioxide emissions when coal is used as a raw material (Tomita, 1991).

Generating method (with coal as a raw material)	Net thermal efficiency (%)	CO_2 emissions (kgC/kwh)
Pulverized coal-fired power plant (China)	28	−0.4
Pulverized coal-fired power plant (Isogo, Kanagawa prefecture, Japan)	36	0.31
Pulverized coal-fired power plant (Matsuura, Nagasaki prefecture, Japan)	38	0.28
Pressurized fluidized bed	42	0.25
Integrated gasification combined cycle	44	0.23
Gasification/molten carbonate fuel cell power plant	50	0.20

*A combined cycle is one in which the heat that is released at the exit of the gas turbine is reused by the steam turbine, thereby increasing the efficiency.

operation. Although this has not yet proceeded to the stage of experiments using a combined cycle, a pilot plant using coal gasification as part of a combined cycle (an integrated gasification combined cycle, IGCC) has reached this phase.

Combined cycles using coal gasification, however, are more difficult to construct than natural gas plants because the coal gas contains materials which are harmful to gas turbines, such as soot, the corrosive vapors of alkaline elements, and hydrogen sulfide etc. Although an obvious point, in order to efficiently collect the energy, it is necessary to maximize the temperature of the gas turbine.

The process first involves feeding a large quantity of limestone into the combustion chamber. At the same time as combustion takes place, desulfurization is performed within the furnace using pressurized fluidized bed combustion. In this technique, the efficiency is increased by employing pressures even greater than in conventional fluidized bed combustion;* in Japan this technology is now at the pilot plant stage.** A topping cycle has also reached the planning stage; in this, partial gasification is performed prior to fluidized bed combustion, and a gas turbine is used in conjunction with the steam turbine; it is expected that this will lead to an efficiency equal to that of the IGCC method but without the technological difficulties associated with the latter.

A fuel cell that will further increase efficiency is currently under development. This directly converts the chemical energy (the availability of which is basically of the same order as that of electricity) into electrical energy; this is achieved without passing through the intermediate stage of low-temperature heat (which is less 'available') or mechanical energy which is easily converted into heat. Development is continuing into techniques that will allow this to be performed at higher temperatures and in conjunction with other procedures such as coal gasification. Table 4.1 shows the expected efficiencies of these processes when coal is employed. Other future technologies with the potential to increase the efficiency of electricity generation include MHD (magnetohydrodynamic) power generation, which is a method of directly generating electricity without using turbines.

Although not restricted to just the generation of electricity, experiments have also been performed to reduce the thermal energy that is removed with nitrogen in the exhaust gases; these employ oxygen-enriched combustion (i.e. combustion conducted under a high pressure of oxygen). However, problems still surround the amount of energy consumed in the production of oxygen and the development of heat-resistant materials. It is also possible to reduce the temperature of combustion by diluting the oxygen using recovered and recycled carbon dioxide.

---

*A stream of air is passed into the bottom of equipment containing small coal particles, causing the coal to rise gently. Under these conditions coal acts in exactly the same manner as water; this is termed a fluidized bed.

**Before being put into commercial operation, tests are performed using a smaller-scale facility in order to determine whether any problems remain. Such a facility is termed a pilot plant.

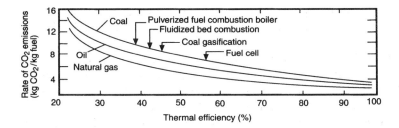

FIGURE 4.2 Thermal efficiencies and CO_2 emissions in power plants using various fossil fuels as raw materials [Hirato, 1991].

In this technique, the carbon dioxide which is produced is not diluted with nitrogen, and consequently it is easy to collect the carbon dioxide as long as the water component is separated; this combustion technique is one that may be suitable for use in the carbon dioxide recovery system described in Chapter 5.

Figure 4.2 shows the relationship between cycle thermal efficiency and the rate of carbon dioxide emissions for coal, oil and natural gas. No matter how great the cycle thermal efficiency, it is impossible for coal to have a lower rate of carbon dioxide emissions than natural gas. However, as was mentioned earlier, if the problem of resources makes it impossible to avoid the use of coal with a high carbon content, then it follows that increasing the cycle thermal efficiency of coal conversion is the best means of restricting the emissions of carbon dioxide.

Accordingly, our objective must be to achieve this goal by carrying out research on coal technology (which is also involved with the problem of carbon dioxide). However, it must be remembered that this is not by any stretch of the imagination a form of nostalgia for coal. It is fervently hoped that it will be possible to increase the use of natural forms of energy such as solar energy and to develop completely safe nuclear power (and nuclear fusion in particular). We coal technologists await the day when our research becomes redundant.

4.3 THE USE OF SECONDARY ENERGY AS A MEANS OF TRANSPORTING PRIMARY ENERGY

When electricity is generated from fossil fuels, it is best to transport the fossil fuel relatively close to the site of consumption and to generate the electricity there. It goes without saying that the same applies to nuclear fuel, which contains a much greater amount of energy per volume than fossil fuels.

However, it is not possible to transport the sun. Because of the locations in which natural forms of energy are found, it is also necessary to develop technology for transporting this energy. Therefore, let us now consider the case of photovoltaic power generation in desert areas. Research in this field has only just begun, but it is hoped that it will lead to the emergence of a new system which will be suited to developments in the primary energy field.

Several methods of transporting energy over large distances are currently under investigation. An example of the costs involved in transportation are shown in Fig. 4.3. The cost involved when the distance is zero reflects the cost of changing the physical form of the energy at the time of transportation (e.g. liquefaction), or the cost of altering the voltage. It does not include the cost of conversion into a different type of energy, such as the conversion of electricity produced in the desert into methanol. At this time, in addition to the cost, the energy efficiency is also important. Bearing this in mind, let us now proceed with the discussion.

Let us consider what would happen if electricity were generated at various sites during daylight using solar energy, and this electricity then transmitted via cables around the world; if this were simply the transmission of energy, the cost would rise steadily with an increase in distance. Of course there would be costs involved with cable construction, but it would be beneficial for the whole human race if cables could extend to those areas which currently have none. However, not all regions have sunlight at the same time; when one has sunshine, others

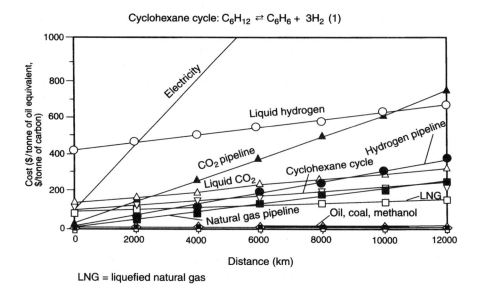

FIGURE 4.3 Costs of transporting various forms of energy and carbon dioxide over long distances [Fujii, 1993].

experience cloud, rain or night. Consequently, this electricity would have to be stored (which is both difficult and expensive) at any individual site of solar power generation unless a system could be established whereby a balance of supply and demand could be achieved across the earth. However, even if this could be done, complete dependence on solar energy would mean that the electricity required at night at one location would have to be transmitted over great distances from other locations which were simultaneously experiencing daylight conditions. Unfortunately, this would result in vast losses of energy.

There are virtually no costs associated with the transportation of materials which can be transported at normal temperatures, such as coal, oil and methanol. Pipelines for the transportation of substances in gaseous form require construction and maintenance. Liquid hydrogen and liquefied natural gas (LNG) require cooling in order to achieve liquefaction; a fixed cost is therefore involved even if the distance is short. In the case of electricity, transformers are also necessary.

The simplest form of energy to produce from electricity is hydrogen; this is possible as long as water is available. Fresh water is available even in deserts; salt water can be obtained even more easily. The energy required to decompose 1 mole of water is a massive 286 kilojoules, and therefore it is only necessary to have a small quantity of water. Provided transportation is over a relatively short distance, the construction of a hydrogen pipeline may be feasible. In actual fact, a large-scale experiment is being performed in Germany in which electricity generated in the African desert is transported to Germany in the form of hydrogen. However, various problems remain concerning the method of later using that hydrogen. Should the energy be converted back into electricity using a fuel cell? Or should it be supplied in gaseous form without undergoing conversion? Should cars run on hydrogen fuel? Are there any safety concerns? Etc., etc. These are just some of the numerous points that cause misgivings.

Another method involves transportation in the form of ammonia. This has the advantage that, simply by the combination of hydrogen with atmospheric nitrogen, a liquid is formed more easily than in the case of hydrogen alone. Also, nitrogen is available everywhere. However, problems still remain in the form of safety worries and corrosion.

Figure 4.3 includes a reference to the cyclohexane cycle. Cyclohexane is a compound which consists of twelve hydrogen atoms attached to a benzene ring made up of six carbon atoms; benzene itself contains only six carbon atoms (C_6H_6). The use of the so-called cyclohexane cycle for transporting hydrogen involves an exchange between these two compounds, i.e. cyclohexane and benzene (Eq. (1)) (it should be noted that approximately the same situation also applies to the methyl compounds of benzene and cyclohexane). In Fig. 4.3, this is the only case that includes the cost of changing the form of the hydrogen.

It should be pointed out that, except for the use of fossil fuels (which are forms of primary energy that exist naturally), the expenditure needed to convert the primary energy into secondary energy must be added to the costs shown in Fig. 4.3. Furthermore, after transportation has been completed, a major additional problem concerns how that energy is used.

In addition to the approaches shown in Fig. 4.3, one view holds that materials should be made in the desert which would allow the electricity to be directly extracted (as in a battery), and that it should be these materials that are transported to the site of consumption. In all cases, the totals of both the energy needed for transportation and also the energy loss involved when re-converting to electricity need to be small. Table 4.2 shows the energy concentrations when these energy media are transported by tanker; a comparison of these energy concentrations (measured per kilogram in the case of materials with a specific gravity greater than one, and per liter for materials with a specific gravity less than one) reveals that only a very few materials can transport energy with approximately the same efficiency as oil (including lithium, iron, aluminum and silicon). However, lithium resources are limited; in the case of iron, aluminum and silicon etc. there are problems concerning the fact that the actual amount of energy required for refining (in this case in order to reduce the oxide to the metal) is much greater than the theoretical value. No matter how much energy is used, the energy extracted in the form of electricity is always less than the theoretical amount.

Table 4.2 Relative efficiencies of various forms of energy transportation (Sano, 1992).

Material	Molar heating value (kcal/mol)	Energy concentration		Molecular weight	Density (g/ml)
		(kcal/kg)	(kcal/l)		
H_2 (gas)	58	(29,000)	2.6	2	9×10^{-5}
Liquid H_2	58	(29,000)	2,010	2	0.07
Li	110	(15,850)	8,460	6.94	0.534
C	94	7,830	(13,300)	12	1.7
Na	96	(4,170)	4,050	23	0.97
Mg	144	5,920	(10,300)	24.3	1.74
Al	200	7,400	(20,000)	27	2.7
Si	209	7,440	(17,500)	28.1	2.35
Ca	152	3,800	(5,900)	40.1	1.55
$Fe^{2.666+}$	89	1,590	(12,500)	55.8	7.87
Fe^{3+}	98	1,750	(13,700)	55.8	7.87
Zn	84	1,280	(9,100)	65.4	7.13
Oil (for reference)	—	(10,000)	9,000	—	—

Note: The figures in parentheses represent factors which have little effect on transportation costs.

4.4 ENERGY TRANSPORTATION BY MEANS OF A CARBON DIOXIDE/METHANOL CYCLE

Finally, let us consider transportation systems involving methanol, which are currently attracting the most attention and investigation in Japan. Recent advances in technology suggest that organic materials would be the easiest to use for transportation. Unfortunately, though, there are no carbon resources in the desert. Thus the plan envisages the liquefied carbon dioxide being transported from the site of consumption back to the desert, where energy would be used to convert the carbon dioxide back into fuel; this fuel would then again be transported to the consumption site and the process repeated. This explains the cost shown in Fig. 4.3 for the transportation of carbon dioxide, which contains absolutely no energy at all.

A proposed system for a methanol/carbon dioxide cycle is shown in Fig. 4.4. Solar energy in the desert is used to convert water into hydrogen, which is then further converted into methanol using carbon dioxide which is shipped in. The methanol is then transported and used for generating electricity; the carbon dioxide which is released is recovered and transported back to the desert in liquid form. Opinion is split as to whether the oxygen which is also produced as a result of the electrolysis of water should be liquefied and returned to the site of consumption. If the methanol is combusted under an atmosphere of oxygen, then as stated earlier it would be simple to recover the carbon dioxide. However, even if partial recovery were performed at the consumption site, the liquefaction of oxygen would require the expenditure of considerable amounts of energy. In addition, it would be necessary to develop combustion technology for oxygen environments, or to recycle the carbon dioxide in order to dilute the oxygen. On the other hand, if combustion were not conducted under an oxygen atmosphere,

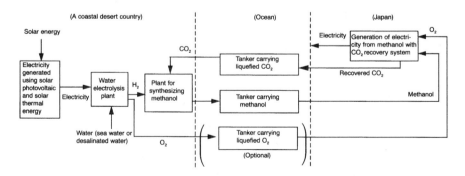

FIGURE 4.4 Transportation of solar energy using a methanol/carbon dioxide system [Park, 1993].

a large amount of energy would be needed to isolate and recover the carbon dioxide contained in the flue gases (this will be discussed later). The research performed to date is insufficient to allow a conclusion to be drawn as to which of these approaches is better.

One point which must not be misconstrued is that this system is definitely not a chemical fixation of carbon dioxide. It is simply the use of carbon dioxide as an energy medium. Of course it certainly does in part involve the recovery of carbon dioxide, which is a topic dealt with in the next chapter, and of course the recovery of carbon dioxide is necessary even in chemical fixation. However, this process is nothing more than a secondary energy system, and from the point of carbon dioxide fixation it would be meaningless to just use carbon dioxide itself. Therefore, it is necessary to correctly evaluate this system from the point of view of primary energy use.

In actual fact, in Japan a national project involving the conversion of carbon dioxide into methanol began with a study of the chemical fixation of carbon dioxide; later, however, the goal of the project was altered and merely became an investigation into the use of carbon dioxide as a transportation medium. It is often said that when a compromise is made and a "new" objective assigned to a project after the project has been initiated, there is a tendency for that project to be evaluated rather leniently. This makes one wonder about whether this is one such instance.

4.5 ENERGY SYSTEMS THAT CONSIDER THE LOCATIONS OF ENERGY SUPPLIES

It is necessary to evaluate secondary energy systems, including the development of hydrogen energy and energy transportation, in terms of technology that systematically and efficiently uses primary energy. The development of secondary energy must then take place in tandem with the development of primary energy sources. It is therefore necessary to perform an overall evaluation as a total system that incorporates all the following steps: primary energy; the transportation of primary energy; the conversion to secondary energy; the transportation of secondary energy; the conversion to the final energy form; and final consumption.

It has to be admitted that such an evaluation is extremely difficult to perform for a number of reasons. First, there is a lack of experience with the transportation of energy other than as primary energy and electricity. Furthermore, the transportation itself is based on the assumption that primary energy will be converted into secondary energy. Second, the end consumer wants energy that is easy to use rather than cheap. On top of that, little consideration is given to factors such as the efficiency of energy use. Even if the energy is rather expensive, the end consumer wastes it without thinking anything of it. And once a person has bought a gasoline-operated car, the car cannot be used as an electric one. Once a business enterprise has made a financial investment, it will not readily switch to another form of energy use until all costs have been recovered. If it is impossible to achieve a return on the investment for several years, then no matter how great the efficiency of the proposed new form of energy, the initial investment will simply not be made. Thus even if a highly efficient system were available, it would be extremely difficult for this to be adopted everywhere.

However, a slightly further development of this argument would suggest that the present sites of energy consumption are not in the best locations; this is a problem related to the siting of factories and the construction of cities. Section 3 of this chapter mentioned to the use of iron as an energy medium; this simply means that the smelting of iron takes place in those places where energy can be produced. If industries which consume a large amount of energy were to move to those places where energy is produced, it would eliminate the need to transport that particular energy. This concept therefore involves a fundamental change in the way of thinking. In the case of cities, people who need information and goods come together, and these people use large quantities of energy. The quantity of energy used for everyday life and for transportation far exceeds the amount used for manufacturing. On the other hand, the situation is developing whereby information can be easily obtained anywhere. Times are changing. The day may come when cities are located on the basis of energy supplies; this would mean that people will live in the desert. However, this would only happen when the transportation of secondary energy is problematical, or when the efficiency of transporting expensive forms of energy is low. This situation could perhaps be regarded as the ultimate secondary energy system for humans.

4.6 ENERGY FOR TRANSPORTATION

It might seem that part of the discussion so far has been rather unrealistic, with the carbon dioxide problem being treated under the supposition that a complex system is far removed from the energy system currently in use. This creates a perplexing situation in which it is difficult to know which energy systems should be considered.

In the previously mentioned methanol system, absolutely no consideration was given to the use of fossil fuels in the generation of electricity. However, what about the energy used for transportation (i.e. the energy used instead of gasoline)? Does it infer that cars could no longer be used?

It is difficult to define the efficiency of the energy used by cars themselves. Of course, before the cars are actually used on the roads, energy is required for materials and manufacture; this, though, amounts to only approximately 20% of the total. Therefore let us assume that the majority of the energy is expended while the cars are being driven. In order to increase the efficiency of energy use, measures are of course needed to increase miniaturization, to reduce wind resistance, and to recover energy during deceleration; here, though, let us simply examine the efficiency of energy use by the engine.

Gasoline engines are estimated to have an efficiency of 20–30%, whereas the most efficient types of diesel engine (those that employ a direct injection system) have an efficiency of approximately 40%. Unfortunately, this does not mean that the use of diesel engines should necessarily be increased. Diesel engines emit large quantities of nitrogen oxides and soot, and it is extremely difficult to solve the problems of both at the same time.

What form should a new system of transportation take? First, a choice must be made between transportation by truck and by rail. Rail transportation is rather inflexible, but requires only a small fraction of the energy needed by trucks. It is thus expected that greater use will be made of containers and so-called "piggy-back" systems (in which each truck will itself be carried by rail).

Research is being carried out into cars that will run on natural gas, hydrogen, methanol, and electricity. However, as mentioned earlier, this author believes that fuel conversion itself (i.e. the use of low-carbon fossil fuels as primary energy) is meaningless in the present situation; if this argument is accepted, the question becomes one of which uses energy in the most efficient way. There are many unclear elements involved in this, and it is necessary to await future technological innovations. However, from the point of view of global warming, the leakage of just 1% of the natural gas would negate any slight increase in efficiency.

If primary energy were to be used in the form of fossil fuels and if oil supplies were to dry up, a compromise would probably be the use of oil produced by coal liquefaction, electricity and methanol. However, when bearing in mind the various environmental problems involved, electricity seems the most attractive option. One possibility would be to use electricity in conjunction with the installation of photovoltaic cells on the car roof. But, as will be mentioned later, the development of electric cars (and especially the capacity of storage cells) is at present facing a serious bottleneck.

The use of hydrogen in everyday life is fraught with danger. Also, as mentioned in the last section, if solar energy in the desert were converted into secondary energy, and then the hydrogen were transported directly from the desert (e.g. to Japan) for use as a fuel for cars, the transportation costs would be prohibitive. This suggests that methanol will perhaps be the fuel that is used in the future.

Figure 4.5 shows the efficiencies of various types of engine. Even the simple replacement of gasoline by methanol would lead to an increase in efficiency and, although not shown in the figure, the same approach would also be applicable to diesel engines. It is said that the replacement of light oil by methanol would permit a reduction in the emissions of nitrogen oxides and soot. Work is in progress to

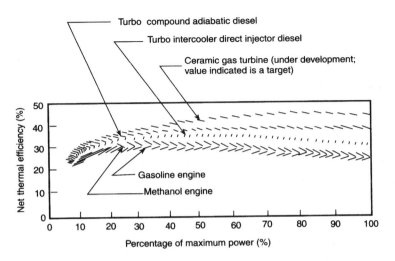

FIGURE 4.5 Thermal efficiencies of various types of engine [Kim, 1990].

develop a ceramic gas turbine as a future engine, as this is expected to be highly efficient.

Two aspects related to electric cars are noteworthy. One concerns the problem of efficiency. It is expected that the efficiency of converting fossil fuels into electricity will one day be between 40% and nearly 50%. On the other hand, the efficiency of converting electricity into a car's mechanical energy is expected to be raised to close to 100%. This therefore means that it is possible to raise the overall efficiency. However, there are problems concerning the storage of electricity and the charging of the vehicles, and in the final analysis, solutions depend on whether suitable battery materials can be developed.

The other aspect relating to electric cars is that the production of carbon dioxide is concentrated at the site of electricity generation since the cars themselves do not release any carbon dioxide (a point which is discussed further in Chapter 5). Specifically, the question is whether the carbon dioxide released at the electricity generating station or the hydrogen manufacturing plant can be recovered and disposed of (a problem also likely to arise with hydrogen-operated cars).

An increase in the efficiency of energy use in transportation becomes more difficult to achieve as the size of the system is reduced. However, that is something that may be resolved by technological innovation. It is also necessary to develop technology which, while improving efficiency, can at the same time curb the emissions of both nitrogen oxides and soot. This is another factor which makes electric cars look a promising choice for the future.

As occurs with cars using gasoline and diesel oil, the recovery of carbon dioxide from the exhaust gases of cars running on methanol is not a very unrealistic proposition. On the other hand, with the methanol system outlined in the Section 4 the energy needed for transportation (i.e. the carbon dioxide emitted by the cars) cannot be supplied to a methanol synthesis system in the desert, and in the end this must be compensated for by the use of fossil fuels. It would also be possible to combine the use of both fossil fuels and the methanol system into a combined system during a transitional stage.

It is also possible to conceive of coal gasification being performed in the desert in the same manner as the system for smelting iron in the desert, which was mentioned in the previous section. If coal were the only source of methanol, there would be a shortage of hydrogen. Therefore this hydrogen shortfall would have to be offset by the production of hydrogen by the electrolysis of water. The electrical energy required for such ancillary operations etc. would be supplied by photovoltaic power generation.

This author is sceptical about the liquefaction of coal, but there is a proposal to use photovoltaic energy to supply the hydrogen necessary for liquefaction. Instead of gasoline, the oil produced from coal liquefaction would act as the energy for transportation.

At the present time, all that can be said is that the future form of cars seems to be completely in the lap of the gods.

4.7   ENERGY CONSERVATION AND HIGH-EFFICIENCY ENERGY USE IN FACTORIES

The reason energy is not used in a highly efficient manner in factories is that this would require companies to make an enormous financial investment, or else because the financial return on the initial investment does not materialize for a considerable period of time.

A further problem concerns the quality and use of the energy. Let us now consider cogeneration, which is a system that employs both electricity and heat. It makes use of the heat produced during the generation of electricity but which at present merely goes to waste. Although practical use is made of the previously unused energy, part of the electricity is naturally sacrificed as a result of the measures which are taken to produce the heat in a form that is easy to supply. When a secondary energy supply system goes as far as including a consideration of thermal energy, cogeneration is a system that can maximize the efficiency of energy conversion. In the final analysis, however, this refers to the total of energy supplied from electricity plus heat, and does not include a consideration of the quality of the energy. In terms of quality, electricity is by far the more available. It is thus necessary to devise a system of energy use that will take full advantage of a cogeneration supply system and which will also adopt an approach that raises the efficiency of both the supply system and the manner in which the energy is used.

As is evident from the example of cogeneration, high-grade energy (e.g. electricity) and low-grade energy (e.g. lukewarm water) may both possess the same amount of energy, but the value of the two is different. Exergy (a measure of available energy) is used as an indicator for evaluating these (including low temperature sources). Certainly, in comparison with the United States, the former Soviet Union and China etc., Japan has made substantial advances in energy conservation, but it is still possible to aim at greater energy conservation by systematically combining the use of various forms of energy in descending order of exergie values (i.e. the use of cascades).

However, in order to promote greater efficiency in energy use, greater energy conservation, and (as discussed in the previous chapter) greater use of unused energy, it is necessary either for energy to become more expensive, or else to introduce an energy tax. Subsidies to support measures for energy conservation have already been instituted, but there should possibly be more money made available for this. Certainly, factories will not implement energy conservation measures unless there is some benefit.

At present, Japan is said to be the number one country in the world with respect to the efficient use of energy, and the most advanced in terms of energy conservation. This situation has come about because at the time of the oil shock Japan's domestic production of oil was negligible, and the country was stunned by the effects of the sudden leap in the price of oil. If it had not been for that, it is unlikely that anyone would have made the investments that they did. If business managers had foreseen the present energy glut, they would not have poured funds into this area. However, the very fact that the investments were made at that time has led to the development of technology which Japan can be proud of. A misfortune was turned into a blessing. Thus, from a long-term perspective, a sudden jump in energy prices is something to be welcomed.

4.8 ENERGY USE AT HOME

Figure 4.6 shows the consumption of energy at home (excluding car use) by country and category of use. In advanced countries, the majority of the energy is used for heating. It certainly would appear from Fig. 4.6 that the household consumption of energy in Japan is low, but this lower figure for Japan merely reflects the fact that the country is helped by having a warm climate. In actual fact, energy consumption in Hokkaido (the northernmost island in the Japanese archipelago, which is approximately at the same latitude as southern France) is roughly the same as in other advanced countries (except for the United States). However, the contribution of Japan's unique form of heating known as a *kotatsu* cannot be overlooked. (A *kotatsu* is a foot-warmer which is usually covered with

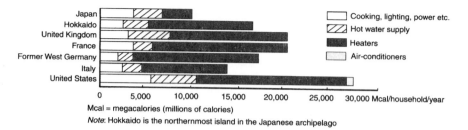

FIGURE 4.6 Household energy consumption by country and type of use [Nakagami, 1990].

a quilt to keep the legs warm.) If the switch to a more Western style of living continues to proceed, the amount of energy needed for heating is likely to increase further. With respect to cooling, the use of air-conditioners has gradually become widespread in America; in Japan, which has relatively hot summers, the use of air-conditioners is likely to increase. When that occurs, it is expected that the spread of households which conserve energy (the construction of which has already been vigorously pursued in places such as Scandinavia) will have a major impact on energy conservation.

The term "energy-efficient households" is an exaggeration, but it refers to the various actions that can be taken, from the hanging of curtains to the construction of walls with materials that provide thermal insulation (such as glass wool), and double or triple glazing for windows. In passing, it should be mentioned that the incorporation of 50 mm of glass wool can lead to energy savings of up to about 40% when rooms are heated. With the exception of Hokkaido, very few energy-efficient homes have so far been built in Japan (only about 15% of homes nationwide). That means that at present the energy savings only come to 6%. That is, the introduction of future conservation measures could potentially lead to the saving of a further 34% of the heating energy.

Even if savings are sought in heating and air-conditioning, the sudden leap in the prices of land and house construction in Japan means that people simply do not have the cash to pay out for the installation of insulating materials. A system to provide interest-free loans for such purposes should be instigated.

4.9 THE ADOPTION OF ENERGY CONSERVATION MEASURES IN OUR DAILY LIVES

On a rather lighter note, let us see whether we use too much energy in our daily lives. When I was a child, no matter how hot the summer, I used to study with my feet placed in ice cold water. In winter I kept warm using a *kotatsu*. I did not ride in taxis; indeed, my family did not even have their own car. Since then, life has become more convenient, until we have arrived at the present situation in which we consume vast amounts of energy. Is it really possible to extricate ourselves from this culture of energy extravagance by increasing efficiency and adopting measures to conserve energy? Does it mean that we have to make sweeping changes in our lifestyle?

Takarada (1993) carried out a survey on eight Japanese university students in order to determine how much carbon dioxide emissions they were responsible for, and to see how much the students could reduce those by themselves. Figure 4.7 shows the amount of carbon dioxide released by each of them between April and June, and Fig. 4.8 gives the amounts released by each type of activity during their daily lives. Two of these students commuted to university from their family homes, and the totals of the emissions from the home were divided by the number of members of the family concerned. All the other six students lived alone. A mere glance at the gasoline consumption in Fig. 4.8 shows which students owned a car.

Figure 4.8 shows clearly that large emissions resulted from the use of baths and cars. Air-conditioners also caused considerable emissions, and it would be expected that the contribution of heaters in winter would likewise be large. The amount of carbon dioxide produced also varied greatly with the method of cooking and boiling water. A microwave oven produced nearly three times as much carbon dioxide as boiling water in an electrical thermos flask; the use of gas also produced much more carbon dioxide than an electrical thermos flask. This latter fact shows that the heating efficiency of gas appliances is surprisingly low, although of course the efficiency of electricity generation is included in these calculations.

Based on these results, the students implemented energy conservation measures which led to a decrease in consumption of 20–60% during the three-month period from July to September (Fig. 4.9). With the elimination of TV watching and driving, the level of stress rose, but there was an increase in the time devoted to reading. Students found that going to bed early and getting up early was if anything refreshing, and certainly not just a negative experience. All the students said that the new lifestyle was hard, but added that they would be able to manage in that way if it were to help the environment. It would be interesting to know how the students are conducting their lives now that the experiment is over.

The researcher in question pointed out that many countries in the world manage to live on only 1% of the energy consumed by advanced nations. We therefore ought to have the leeway to test out how we can cut down on our energy use.

Finally, what kind of system of energy use should the developing world adopt? Surely it must not be like the egotistical approach of the advanced countries but rather one that considers the effects on the entire planet.

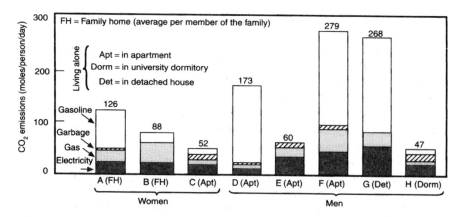

FIGURE 4.7 Carbon dioxide emissions per student (before implementation of conservation measures) [Takarada, 1993].

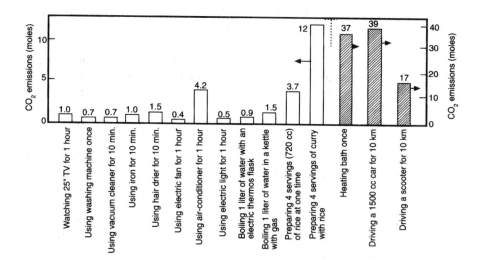

FIGURE 4.8 Carbon dioxide emissions during various daily activities [Takarada, 1993].

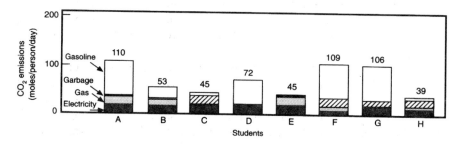

FIGURE 4.9 Carbon dioxide emissions per student (after implementation of conservation measures) [Takarada, 1993].

References

Fujii, Y. 1993. Evaluation of countermeasures against carbon dioxide in energy systems (doctoral thesis). The University of Tokyo, Japan (in Japanese)

Hirato, M. 1991. *Coal Utilization from the Viewpoint of the Carbon Dioxide Problem*. In Tomita, A. (ed.). Report of Grant-in-Aid for Scientific Research, The Ministry of Education Science and Culture, Japan (in Japanese)

Kim, H. 1990. In Komiyama, H. *et al.* (eds.). 1990. A Handbook of the Global Warming Issue, p. 556, IPC, Tokyo, Japan (in Japanese)

Nakagami, H. 1990. In Komiyama, H. *et al.* (eds.). 1990. *A Handbook of the Global Warming Issue*, p. 159, IPC, Tokyo, Japan (in Japanese)

Park, P., Y. Tazaki and Y. Suzuki. 1993. *Energy and Resources* **14**, 72 (in Japanese)

Sano, H. 1992. *Iron and Steel* **78**, 1277 (in Japanese)

Takarada, T. 1993. *Kagaku Kogaku* **57**, 72 (in Japanese)

Tokyo Electric Power Company. 1992. *Report on "Action Plan for Environment"* Tokyo, Japan

Tomita, A. 1991. In Tomita, A. (ed.). Coal Utilization from the Viewpoint of the Carbon Dioxide Problem. Report of Grant-in-Aid for Scientific Research, The Ministry of Education Science and Culture, Japan (in Japanese)

5 TECHNOLOGIES FOR THE RECOVERY, STORAGE, AND UTILIZATION OF CARBON DIOXIDE EMITTED AS A RESULT OF ENERGY USE

5.1 TECHNOLOGY FOR THE RECOVERY OF CARBON DIOXIDE

Techniques for recovering carbon dioxide basically fall into the following three categories:

(1) *Accelerated absorption of atmospheric carbon dioxide by natural mechanisms.* This is primarily concerned with plant life and ocean environments; the role played by these in the earth's carbon cycle will be discussed in Chapters 6 and 7.

(2) *Recovery and isolation (storage) of carbon dioxide from high-emission sources.* A summary of isolation techniques employing the oceans will also be included in this chapter since the oceans are a possible storage site for the carbon dioxide that originates from high-emission sources, although a full discussion of the principles involved will be deferred until Chapter 7.

(3) *Chemical or biological conversion of recovered carbon dioxide, and its subsequent fixation and/or utilization.* This has virtually no relevance to a discussion of the carbon dioxide problem from an energy viewpoint; it will be repeatedly pointed out that an evaluation of such recovery technology needs to be carried out from a different perspective.

First there will be a consideration of category #3, i.e. the conversion of carbon dioxide into (and subsequent fixation as) organic chemicals by utilizing the growth of marine microorganisms and vegetation such as algae which participate in photosynthesis. There will also be a discussion of related geochemical principles, followed by an evaluation of techniques for inorganic chemical conversion and fixation. Conversion technology using coral will only be considered from the standpoint of inorganic chemistry. The question of whether these technologies actually result in the fixation of carbon dioxide will also be addressed.

Following that will be a discussion of category #2 (i.e. the recovery and isolation from high-emission sources) in which various techniques of sequestering recovered carbon dioxide will be reviewed (i.e. ocean disposal, enhanced oil recovery, and storage in depleted natural gas wells and aquifers).

The techniques are discussed on the premise that it is possible either for the carbon dioxide to be fixed, and then stored (or even utilized), or else that the carbon dioxide can be recovered and used as a medium for the transportation of energy (as outlined in Chapter 4). The techniques referred to are methods for the recovery of carbon dioxide from flue gases; processes involved in the production of energy from fossil fuels (under the assumption that any carbon dioxide produced will subsequently be recovered); and the centralization of the sources of carbon dioxide emissions.

5.2   CARBON DIOXIDE: RECOVERY AND FIXATION, OR RECYCLING VIA ORGANIC CHEMICALS

There have been suggestions that carbon dioxide should be recycled via organic chemicals, which involves the conversion of carbon dioxide into useful organic materials. Proposed methods for accomplishing this are catalytic hydrogenation (Fig. 5.1), and electrochemical and photochemical reduction (Figs. 5.2 and 5.3). In each case, the organic materials synthesized by the process would be useful in, but would again lead to the formation of carbon dioxide. When the fuel is combusted, carbon dioxide is produced and energy is released; this means that at the

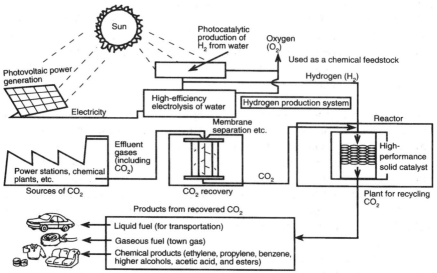

FIGURE 5.1   Reuse of carbon dioxide by catalytic hydrogenation [Arakawa, 1990].

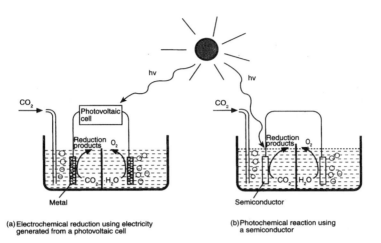

FIGURE 5.2 Photochemical and electrochemical reduction of carbon dioxide [Fujishima, 1993].

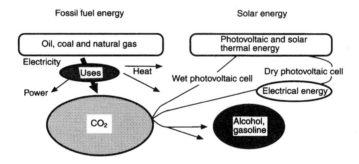

FIGURE 5.3 Reuse of carbon dioxide as an energy resource using photochemical and electrochemical techniques [Fujishima, 1993].

time of resource regeneration, an amount of energy must be provided as input which is equal to at least the amount of energy released. Unless some renewable form of energy such as solar energy is employed, carbon dioxide would have to be produced somewhere in order to replace the released energy (this is the same situation as occurs in the manufacture of materials that are hard to decompose, such as polymers). Thus renewable energy sources are essential to the process. Photochemical reduction makes direct use of solar radiation; electrolytic reduction utilizes electricity generated from solar radiation; and catalytic hydrogenation employs the hydrogen which is produced by that electricity as an energy source. This argument assumes that the recovered carbon dioxide originally comes from fossil fuels or biomass. However, the recovery of carbon dioxide from the atmosphere is not feasible since too much energy would be required.

To put it another way, when a useful product is manufactured, the simplest method of production is also the best from the energy standpoint. The use of carbon dioxide as a raw material might appear attractive if this were to lead to the manufacture of a better product or if the whole process were to result in energy conservation. Generally, however, it would be unprofitable to recover and reuse carbon dioxide, which has the lowest energy content of any carbonaceous material. There ought to be a more energy-efficient conversion process than converting the released thermal energy into chemical energy.

Even if a more efficient process does not exist, there is scope for comprehensive studies into determining whether the manufacture of useful products by the catalytic conversion of carbon dioxide using the waste heat from power stations (even if this means utilizing currently unused energy) is actually the best method of energy conversion. In addition, there are various other chemical processes that do not use carbon dioxide as a raw material; however, these utilize previously unused energy and do not regenerate carbon dioxide as an energy resource.

Certainly, it is possible to make a wide range of materials from carbon dioxide by using energy in the form of hydrogen etc. However, since these techniques are basically not concerned with measures to counter the problems caused by carbon dioxide, they will not be discussed here.

The situation is much clearer regarding the reduction of carbon dioxide to carbon and its subsequent fixation. It is absolutely impossible to convert carbon dioxide back to carbon without replacing the energy that was released during the combustion of the original carbonaceous material, although it may not be necessary to replace the total energy possessed by all hydrogen atoms which were present in the fossil fuel. In that case, however, the "Hydrocarb" method described in Section 3.8 (in which only the hydrogen content of the coal.is used, with the carbon content being reburied in the mine) would probably be preferable since carbon dioxide is not formed in the first place. To put this another way, in order to convert the carbon dioxide into a form that can easily be fixed, it is necessary to supply quantities of energy greater than that produced when the carbon dioxide is released from the fossil fuel. From the energy balance standpoint alone, this is not theoretically feasible.

Let us now consider energy media that use carbon-based materials such as methanol. This first involves the carbon dioxide which is emitted being returned from the country in which the energy was consumed back to the country that originally exported the natural primary energy (such as photovoltaic energy). The carbon dioxide is then converted into a transportable energy medium and transported back to the energy-consuming country, where it is again used as a secondary energy source. If this process were the most efficient method of transporting energy from the country with primary energy supplies to the consumer country, it would help to contribute to the resolution of the problems concerning carbon dioxide; however, as stated in Chapter 4, this process does not bring about sequestration, removal or fixation of the carbon dioxide. It is therefore necessary to consider this process as the production and transportation of primary energy (i.e. solar energy). This argument applies not only to hydrogenation and electrolytic reduction, but also to photochemical reduction processes. The process needs to be compared with the direct generation of electricity using photovoltaic cells in conjunction with other methods of energy transportation.

The above proposal cannot be completely disregarded since energy transportation certainly plays a role. However, when such transportation is not necessary, it makes absolutely no sense to first convert solar energy into electricity (which is highly efficient and usable) and then transform it (incurring a partial energy loss) via hydrogen into methanol. In comparison, the reduction of carbon dioxide by hydrogen would involve the recovery of carbon dioxide in a suitably designed large energy facility. This would entail direct transportation of the hydrogen to the facility and the direct use of that hydrogen energy; accordingly, the release of carbon dioxide would have to be prevented, but it would be possible to reduce the fossil fuel used by an amount equivalent to the energy supplied by the hydrogen. In this way, it would be possible to use hydrogen as an energy source in a large facility if the facility were capable of recovering the carbon dioxide produced.

Approximately the same situation exists when other materials are produced or when electrolytic reduction is employed. The photochemical conversion of solar energy into chemical energy needs to be compared with the conversion of solar energy into electricity (using photovoltaic cells or solar thermal power), and the further conversion of that electricity into transportable chemical compounds. Questions that need to be addressed are whether photochemical conversion has a greater conversion efficiency; whether it would be preferable for the evaluation to incorporate the whole process as far as the final consumption stage; and whether photochemical conversion may be more economical since it requires less initial investment. These issues, which naturally assume future advances in technology, need to be debated accurately and without bias. It should also be noted that under such a scenario electricity, hydrogen and other secondary energy systems would become competitors.

At the present time it is still too early to seriously contemplate the construction of energy systems in which carbon is recycled as a useful resource (indeed, such a day is unlikely to dawn). Only a small fraction of the carbon in fossil fuels is actually used as a carbon resource (i.e. for the manufacture of chemicals), and the use of fossil fuels in such small quantities is perhaps permissible. On the other hand, however, for the existence of human beings on the earth's surface, vegetation is indispensable, and its disappearance is inconceivable. It is certain that biomass (which comprises energy that at present goes mostly unused) will be formed as a byproduct in food production. Even if there were no fossil fuels, or if these existed but could not be used, the amount of biomass available should be equal to or greater than the quantity of carbon resources used today as raw materials for the chemical industry.

As justification for the reduction of carbon dioxide, some researchers argue that in the world of nature, solar energy is converted directly into chemical energy, and carbon dioxide is recycled; they therefore claim that it is important to construct manmade systems of photosynthesis. However, in the natural carbon cycle the only products of photosynthesis are food and oxygen—products that are essential for the respiration and activities of animal life (including microorganisms). The process does not involve the recycling of the carbon dioxide that is produced when energy is used as a result of the actions of human beings. It is often said that the Amazon acts as the lungs of the world in the sense that the Amazon absorbs carbon dioxide. However, this is a misnomer; the Amazon cannot resolve the problems stemming from carbon dioxide emissions. Human beings therefore have no option but to construct new energy systems that incorporate the absorption or fixation of carbon dioxide, or new systems of renewable energy that replace fossil fuels. If we are to imitate plant life, the recycling of carbon dioxide can be considered only for food production; it is not an option for the carbon dioxide that is released from fossil fuels.

Let us again briefly re-summarize the situation. There is absolutely no merit in developing technology solely for the purpose of solving the problems associated with carbon dioxide which is based on the use of carbon dioxide as part of an energy system. The use of carbon dioxide must be accurately evaluated in terms of the overall energy system. Likewise in the case of hydrogen energy, the development of primary energy sources for the production of hydrogen has a greater priority than the use of hydrogen energy itself. Alternatively, hydrogen energy should be considered as a secondary energy source for the development of a primary energy source.

Based on the assumption that carbon dioxide will be recovered, isolated and stored (as previously mentioned, hydrogenation is not feasible), it is also possible to carry out a conversion to low-carbon fuels (except hydrogen) for public consumption by utilizing the hydrogen which is obtained from the energy released when the carbon contained in these fuels is converted into carbon dioxide. (It is possible to supply hydrogen directly, but this would not be practical because of the dangers associated with its use in everyday life.) The process of centralizing the emission sources of carbon dioxide is discussed at length in Section 10 of this chapter.

5.3 CARBON FIXATION BY THE GROWTH OF PLANT BIOTA SUCH AS PHYTOPLANKTON AND ALGAE, AND A COMPARISON WITH FIXATION VIA ORGANIC CHEMICALS

Carbon fixation by the growth of organisms can be considered to be approximately the same as the photochemical reduction of carbon dioxide which was discussed in the previous section. That is, phytoplankton cultivated using carbon dioxide and sunlight will in due course decompose with the release of carbon dioxide. Therefore this does not result in the fixation of the carbon dioxide, which needs to be utilized in some way. Even if only a minute fraction is converted into useful materials, the large remaining quantity of biomass can only be used as an energy source (it should be noted that the recovery of carbon dioxide is only feasible if large quantities are involved).

In order to achieve the fixation of carbon dioxide, it is necessary to provide more energy than is obtained when carbon dioxide is released from fossil fuels. Of course, the energy that is added should be in the form of renewable energy such as solar energy etc. Thus the process first involves the combustion of fossil fuels with the release of carbon dioxide. No matter whether the carbon dioxide is released into the atmosphere and then recovered, or whether it is immediately recovered and then directly used, the carbon dioxide is again returned to organic chemicals [Fig. 5.4(a)]; from this point it is similar to the chemical method discussed in the previous section. However, as mentioned earlier, if the products are used as an energy source (including those used in transportation) it will involve the recycling of carbon dioxide as shown in Fig. 5.4(b). This is therefore a secondary energy system that uses solar energy as the primary energy source [Fig. 5.4(c)]; thus it is necessary to compare this system with other types of systems that employ primary energy. It should be clearly realized that in this case also the use of carbon dioxide as an energy medium is in itself pointless.

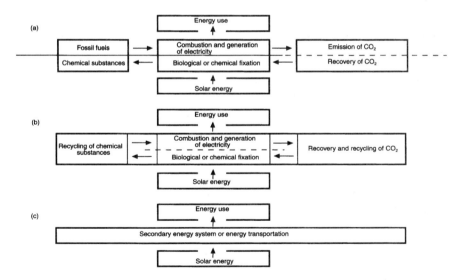

FIGURE 5.4 (a) Biological or chemical fixation, (b) recycling of carbon dioxide, and (c) a secondary energy system [Kojima, 1995].

Based on the above discussion, let us compare electricity generation systems using photovoltaic cells with similar systems that employ biological fixation (electrical systems are the best for comparison since electricity is the most convenient to use). As will be described in Chapter 6, when plants undergo photosynthesis the efficiency of energy conversion is generally estimated to be at most of the order of 1%, which only rises to several per cent when algae are cultivated under a carbon dioxide-rich atmosphere. Whereas the efficiency of coal-to-electricity conversion is 36–40%, in the case of algae the drying processes also require energy; therefore let us assume that the efficiency of conversion into electricity for algae is of the order of 30% when the algae are dried and combusted. This means that the overall efficiency of the process is still a little short of 2%. In comparison, the efficiency of photovoltaic cells easily exceeds 10%. Even if biomass is cultivated and then converted into ethanol for use in transportation (as happens in Brazil) it is expected that the overall efficiency of energy conversion will probably not exceed 5%. Moreover, if the thermal efficiency of an engine is taken to be about 30%, the overall efficiency will be a little less than 2%. In contrast, electric vehicles powered by photovoltaic cells are expected to have a conversion efficiency of approximately 10%.

Since electricity cannot be transported for very long distances, it may be necessary to carry out more accurate evaluations that incorporate an appraisal of issues related to transportation. However, it is not necessary to consider carbon dioxide itself when carrying out an evaluation of energy efficiency.

Another problem surrounds the use of solar energy via the cultivation of marine microorganisms and algae. Specifically, the cultivation of organisms requires fertilizer (especially phosphorus and nitrogen). The phosphorus and nitrogen present in the harvested organisms must be recycled and used again to promote the growth of organisms [Fig. 5.5(a)]. Unfortunately, this technique is currently considered to be extremely difficult to implement.

In marine microorganisms, the ratio of phosphorus to nitrogen and carbon is estimated to be 1:16:106. In order to produce enough phosphorus and nitrogen to fix 5.5 gigatons of carbon equivalent emitted into the atmosphere following the combustion of fossil fuels, it would be necessary to manufacture over ten times the current level of fertilizer production.

It should be noted that if phosphorus and nitrogen fertilizer were to be spread over the ocean (a technique known as ocean fertilization or ocean biomass stimulation), it is believed that the corresponding amount of carbon would be fixed naturally by photosynthesis, with the net result being no overall increase in carbon dioxide released into the atmosphere. This is due to the fact that the oceans are relatively poor in nutrients with the exception of extremely small areas such as coastal regions (for details, see Chapter 7). An outline of this ocean fertilization technique is given in Fig. 5.5(b). If this were successfully put into practice, it would be acceptable to combust fossil fuels and release into the atmosphere an amount of carbon dioxide corresponding to the quantity fixed by this method.

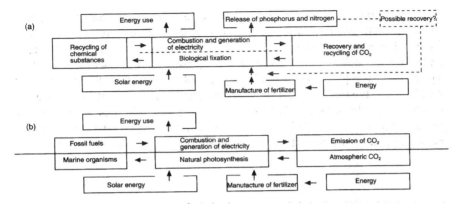

FIGURE 5.5 Comparison of two biological fixation processes: (a) an energy system utilizing microorganisms, and (b) ocean fertilization [Kojima, 1995].

It is necessary to consider whether it is better to use an energy system based on microorganisms [Fig. 5.5(a)] or to use ocean fertilization [Fig. 5.5(b)]. It is doubtful whether the microorganism approach would actually result in energy production. Furthermore, the implementation of the microorganism approach in coastal waters would create a problem in that any surplus fertilizer not used for plant growth would result in the water becoming rich in nutrients, which in turn invites the problems associated with eutrophication (i.e. the proliferation of dense plant growth would mean that animal and fish life would be unable to survive the resulting deprivation of oxygen). If cultivation were instead to be carried out in tanks on land, the cost of land would become a factor and it would become necessary to concentrate the solar energy (possibly by the use of optical fibers).

In contrast, the ocean fertilization approach can be put into practice without difficulty using existing technology. Also, since ocean fertilization would be carried out in nutrient-poor areas of the ocean without the use of manpower the problems associated with the use of microorganisms would not occur.

Since the net quantity of carbon dioxide neither increases or decreases, the microorganism technique cannot be evaluated as a system for isolating and fixing carbon dioxide as has been discussed in this section. It is therefore necessary to evaluate the technique as a system that utilizes solar energy and in which the recovered carbon dioxide is recycled. If this microorganism technique should indeed become feasible in future as the result of technological developments, then improvements in the efficiency of energy conversion (including the production of biomass) would become important, and the development of technology for the recovery and recycling of phosphorus and nitrogen would also be necessary.

Ocean biomass that is obtained in the above way could open up a myriad of possible applications. It could be used as fodder for livestock, which would reduce the amount of grazing land required; this land would then become available for afforestation, which in turn would play a role in carbon fixation. A further advantage is the fact that marine microorganisms contain far more phosphorus and nitrogen than land vegetation (this is because so much phosphorus and nitrogen is needed for the cultivation of these microorganisms). Consequently, since land vegetation only requires between 0.1% and 1% of the amount of fertilizer per unit of carbon, this biomass (or the excrement and urine from the livestock that feed on it) would form the optimum fertilizer. Ocean biomass could thus be used for the greenification of deserts. This demonstrates why it is important to approach the problems associated with carbon dioxide from a multitude of perspectives.

5.4 INORGANIC FIXATION AND THE GEOCHEMICAL CARBON CYCLE

Many non-specialists at first seem to believe that it would be alright if carbon dioxide were to be absorbed by either calcium hydroxide (slaked lime) or calcium oxide (quicklime). However, both of these substances originally existed in the form of limestone. As a result, during the manufacture of cement (which essentially is the calcination of limestone) carbon dioxide is released. Consequently, it is easy to see that such a procedure would have no value in terms of carbon fixation, in the same way that the above-mentioned organic fixation was pointless.

A perusal of the history of the earth would reveal that carbon dioxide was initially present at a pressure of several dozen atmospheres. This carbon dioxide is said to have later undergone fixation by coral. In actual fact, however, it was fixed by the conversion of calcium silicate into calcium carbonate in accordance with Eq. (1). Certainly, carbonates exist in the remains of living things, e.g. in coral reefs:

$$CO_2 + CaSiO_3 = CaCO_3 + SiO_2 \tag{1}$$

$$2CO_2 + CaSiO_3 + H_2O = Ca^{2+} + 2HCO_3^- + SiO_2 \tag{2}$$

$$Ca^{2+} + 2HCO_3^- = CaCO_3 + CO_2 + H_2O \tag{3}$$

$$CaCO_3 + CO_2 + H_2O = Ca^{2+} + 2HCO_3^- . \tag{4}$$

Another silicate, magnesium silicate ($MgSiO_3$), undergoes a similar reaction, producing magnesium carbonate ($MgCO_3$) instead of calcium carbonate ($CaCO_3$). Anorthite ($CaAl_2Si_2O_8$) acts similarly to produce calcium carbonate (lime), and $Al_2Si_2O_5(OH)_4$ instead of silicon dioxide (SiO_2). In each case, carbon dioxide is absorbed when the silicate is converted into the carbonate.

Equation (1) is actually an over-simplification, and can be considered the sum of two separate reactions. First, the weathering process of silicate-bearing rock

[Eq. (2)] occurs mainly on land; after that, the calcium ions, which are present in excess, mainly react in the surface layers of the ocean to form calcium carbonate [Eq. (3)]. In the deep ocean, Eq. (3) proceeds in the reverse direction.

One countermeasure against carbon dioxide involves the use of coral reefs (a process which will be discussed in depth in Chapter 7). From the viewpoint of pure inorganic chemistry, this is a process in which two moles of bicarbonate ions which are dissolved in sea water are converted into solid calcium carbonate and gaseous carbon dioxide [Eq. (3)]. That is, the carbon dioxide which undergoes fixation in solidified form does not actually come from the atmosphere; also, the fixation is accompanied by the release of carbon dioxide into the atmosphere. In other words, it is a countermeasure that produces undesirable results.

Figure 5.6 shows the geochemical carbon cycle, including the above processes. Over an unimaginably long period of time the earth's magma converts calcium carbonate into calcium silicate, with carbon dioxide being re-released as a result of volcanic activity. This cycle repeats itself over several dozen millennia, or even over hundreds of millions of years.

Equation (3) on its own results in the release of carbon dioxide. However, since Eq. (2) results in the absorption of twice the amount of carbon dioxide that is produced by Eq. (3), the net result is a positive absorption [as indicated by Eq. (1), which is a combination of Eqs. (2) and (3)]. Therefore, if the silicates referred to in the equations are artificially made to undergo weathering processes in accordance with Eq. (2), carbon dioxide will of course be absorbed; however, the geochemical cycle will then proceed further, and Eq. (3) will mean that carbon dioxide is again released into the atmosphere, and as a result the effectiveness of the overall process is halved.

$$CO_2 + CaSiO_3 \longrightarrow CaCO_3 + SiO_2 \tag{1}$$

$$2CO_2 + CaSiO_3 + H_2O \longrightarrow Ca^{2+} + 2HCO_3^- + SiO_2 \tag{2}$$

$$Ca^{2+} + 2HCO_3^- \longrightarrow CaCO_3 + CO_2 + H_2O \tag{3}$$

$$CaCO_3 + CO_2 + H_2O \longrightarrow Ca^{2+} + 2HCO_3^- \tag{4}$$

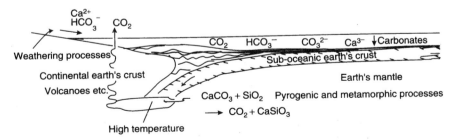

FIGURE 5.6 The geochemical carbon cycle [Ohsumi, 1991].

In concrete terms, silicate-bearing rock is converted into the form of a slurry, into which flue gases (including carbon dioxide) are introduced. The resulting solution is neutral or weakly alkaline; it is thus acceptable to release this into the ocean in that form. It is also acceptable to force it into coral reefs so as to act as a source of calcium ions. Even if coral is then produced by the reaction described by Eq. (3), the overall outcome of the countermeasure would still be positive. Furthermore, the resources are large enough for this technique to be put into operation (Kojima *et al.*, 1996).

However, this still leaves many other questions; for example, how small should the diameter of the particles be? Consideration must also be given to the energy which is required for the crushing, since this leads to the release of carbon dioxide. If it is necessary to absorb highly concentrated carbon dioxide, energy would also be needed for isolating the carbon dioxide.

What would happen if silicate-bearing rock were turned into a slurry, made to absorb carbon dioxide, and then forced into the deep ocean? Unlike in the surface layers, in the deep ocean the equilibrium point of Eq. (3) would mean that the reaction would proceed in the reverse direction, i.e. in the manner described by Eq. (4). Therefore, if it were mixed well with sea water and became dilute, Eq. (3) would not proceed. That is, two moles of carbon dioxide would actually be absorbed. Even in the unlikely case of Eq. (3) taking place, a considerable amount of time would elapse before materials in the deep ocean appeared at the surface. At this time, energy would be required to transport the solution into the deep water; however, if it was converted into a slurry, producing an apparently large density, it is likely that it would descend naturally into the deep ocean.

Weathering is also observed in the case of limestone, with the absorption of one mole of carbon dioxide [Eq. (4)]. Of course, if this rises into the surface ocean layers the reverse reaction occurs [Eq. (3)], and the overall situation is not improved. Let us consider what would happen if the absorbed liquid was injected into the deep ocean in that form, or if, even without Eq. (4) continuing to completion, the material was converted into a slurry and subsequently introduced into the deep ocean. Unlike the case of silicates, only one mole of carbon dioxide would be involved, and carbon dioxide would undergo fixation. It is also unnecessary to consider the possibility of the reverse reaction [i.e. Eq. (3)] occurring in the deep ocean. Furthermore, as is discussed in Chapter 7, this approach is feasible since several centuries would pass before the materials appeared at the surface. Unlike the disposal of liquid carbon dioxide in the deep ocean, there would be no worries regarding the lowering of the pH of sea water.

However, when all is said and done, a great many problems need further investigation, such as whether it is appropriate to use an amount of rock comparable with an excessive quantity of carbon dioxide, and whether there are indeed effects on the marine environment (even though the pH of the sea water would not decrease).

5.5   PROPOSALS FOR INORGANIC FIXATION

Although the proposal described in the previous section is based on the author's own ideas, similar ideas do of course exist and it cannot be claimed to be completely original. This section will introduce proposals that have been put forward by other researchers.

The first proposal concerns transporting sea water to the desert and evaporating off the water content. The difference in solubility would mean that calcium carbonate would be deposited before salt; the salt would then be returned to the ocean in solution. However, when the whole ocean is considered, it needs to be realized that Eq. (3) would occur in the desert. Since the remaining sea water is returned to the ocean, the carbon dioxide produced as a result of Eq. (3) would presumably be released into the atmosphere from somewhere in the oceans even if this did not occur in the desert. This would, though, be cancelled out by the elution of calcium and magnesium due to weathering [Eq. (2)] (Fig. 5.7). The problem concerns how to bring about the reaction described by Eq. (2) while at the same time avoiding the occurrence of Eq. (3). However, Eq. (3) inevitably occurs in the surface waters of the oceans and in the desert, thus rendering this proposal totally unrealistic.

Next, the same researcher (Tanaka, 1993) takes advantage of the fact that the amount of carbonates contained in desert sand is equivalent to 5×10^{18} g of carbon dioxide, or 1,000 gigatons of carbon. This exceeds the amount of organisms shown in Fig. 2.7, and amounts to a total of 2,000 tonnes of carbon per hectare. Since this is nearly ten times the amount of carbon retained per hectare of tropical forest, and since deserts are arid regions, this indicates the existence of a carbon sink. When this amount is multiplied by the recent rate of desertification (6 million hectares per year), the figure actually comes to 12.0 gigatons. This certainly does not mean that this amount is suddenly accumulated in the short space of one year, but isotopic measurements provide an estimated value of the order of several hundred million tonnes per year, and it is inferred that this may be part of a missing sink. Tanaka therefore asserts that the existence of deserts is valuable since the deserts themselves act as a large carbon sink. But even if

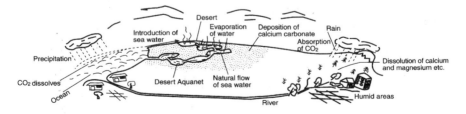

FIGURE 5.7   Fixation of carbon dioxide due to deposition of calcium carbonate in the desert [Tanaka, 1989].

deserts are "destroyed" and replaced by forests, the reverse reaction of that shown in Eq. (1) requires an enormous amount of energy (such as exists in the earth's magma) and it cannot take place at the earth's surface. The only thing that is possible is actual absorption of carbon dioxide by the reaction described by Eq. (4). Furthermore, as will be discussed in the next chapter, it is obvious that if the desert is greenified, carbon will become fixed. Thus the "destruction" of the deserts is a good thing—at least in those areas which have become deserts as a result of human activity.

Next let us consider the absorption of carbon dioxide by sea water, and its subsequent fixation in the form of calcium carbonate and magnesium carbonate. As will be described in the next chapter, it is evident from Eq. (3) that rather than carbon dioxide gas being converted into calcium carbonate and fixed, bicarbonate ions which are present in the sea water will react with calcium ions to form calcium carbonate, with the accompanying release of carbon dioxide from the ocean. As described in the previous section, apparently good methods of fixing atmospheric carbon dioxide as calcium carbonate would be to either use alkaline calcium hydroxide, or to add some other alkaline hydroxide to the calcium in the sea water. This would then lead to the reaction described by Eq. (5):

$$Ca^{2+} + 2OH^- + CO_2 = CaCO_3 + H_2O \tag{5}$$

$$2NaCl + 2H_2O = H_2 + 2NaOH + Cl_2 \tag{6}$$

Figure 5.8 is a process flow diagram for a proposed method of absorbing carbon dioxide in which the hydroxyl group is obtained by electrolysis using electricity generated by photovoltaic cells. The electrolytic reaction is shown by Eq. (6).

$$Ca^{2+} + 2OH^- + CO_2 \rightarrow CaCO_3 + H_2O \tag{5}$$

$$2NaCl + 2H_2O \rightarrow H_2 + 2NaOH + Cl_2 \tag{6}$$

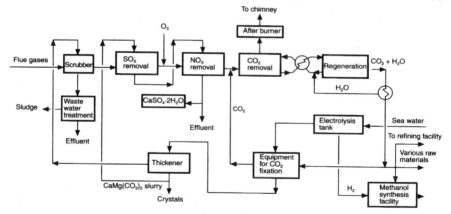

FIGURE 5.8 Proposed method of carbon dioxide absorption [Takeuchi, 1992].

The combination of Eqs. (5) and (6) leads to the production of hydrogen and chlorine. We can assume that the hydrogen is used for the synthesis of methanol or as a source of hydrogen energy. But what about the chlorine? The first thing that comes to mind is use as vinyl chloride, C_2H_3Cl, a raw material for the production of vinyl plastics. In vinyl chloride there are two carbon atoms for every atom of chlorine; treatment of 1 mole of carbon dioxide would produce *two* atoms of chlorine, and therefore the manufacture of 2 moles of C_2H_3Cl would require four atoms of carbon. On the other hand, the majority of fossil fuels are used as sources of energy; in Japan, those fossil fuels used as raw materials for the petrochemical industry merely amount to approximately 1% of crude oil imports. It is thus obvious that not all the carbon dioxide can be absorbed using this system. Of course, it is nonsense to consider the absorption of chlorine by alkalis.

The above discussion demonstrates the vast scale of the carbon dioxide problem. Even the proposer of the above scheme acknowledges that it can only be effectively implemented on a limited scale. Indeed, this author wonders whether the idea of treating the problem within the chemical industry is simply a non-starter.

5.6   OCEAN DISPOSAL

Although the recovery of carbon dioxide from flue gases certainly involves the expenditure of both money and energy, it can be done using existing technology, and it is likely that improvements in efficiency can be achieved (this will be discussed further in Section 8). However, the basic problem concerns how the isolated carbon dioxide should be handled. As was mentioned at the beginning, with the exception of absorption and fixation by silicates, conversion into other chemical substances appears to be difficult. If so, the only remaining solution is to store it somewhere in its existing form. The following discussion considers a proposed example of such storage.

As far as Japan is concerned, the main candidate as a storage site is the ocean. First, since the oceans have an extremely large absorptive capacity, it has been proposed that flue gases be directly absorbed by sea water. However, if the amount of dissolved carbon dioxide greatly increases over the present level, it is feared that sea water will become acidic; in addition, it is necessary to investigate whether it is possible for the carbon dioxide, once it has been dissolved, to be later re-released. If the scale becomes too large, it is likely that the pumping up of sea water on its own will lead to an overall energy loss, even though there is no need to recover the carbon dioxide. This particularly applies to the raising of water from the deep ocean, where the capacity to absorb carbon dioxide is great. Another proposal along these lines considers using the existing downwelling of water.

Figure 5.9 shows the situation in which gaseous carbon dioxide is injected into shallow sea water by means of a pipeline at a location where the density gradient of sea water is not great; this occurs at depths of 200–300 meters. The carbon dioxide dissolves in the sea water, thereby increasing the specific gravity of the water and causing it to sink first to the sea bed, and then along the bed to the deep ocean, where dispersion takes place along a plane of equal density. As is also explained in Chapter 7, it is likely to either be preserved for 1,000 years or else (in the case of the Pacific Ocean) to react with the calcium carbonate which is accumulated at the sea bed [Eq. (4)].

Steinberg *et al.* (1984) performed calculations for a method in which carbon dioxide is recovered from flue gases, pressurized to produce liquefaction, and then injected into the deep ocean (Fig. 5.10). The former state-owned company of British Coal indicated to the author that this particular method was not of interest to them since no deep ocean exists near the United Kingdom (this technique is only feasible at locations near deep ocean waters). In the case of Japan,

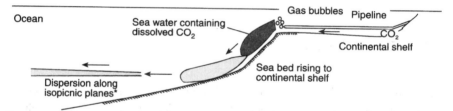

*isopicnic plane = contour of equal density

FIGURE 5.9 Injection of CO_2 into shallow sea via a pipeline [Haugan and Drange, 1992].

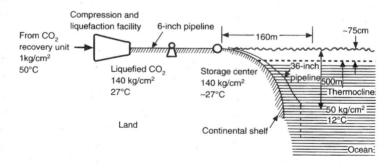

FIGURE 5.10 Steinberg's proposed process for recovery, liquefaction, and ocean disposal of carbon dioxide [Steinberg, 1984].

however, it is only a short distance to the abyssal Japan Deep, and the approach may be of value.

If this technique is applied to the generation of thermoelectric power using fossil fuels, it still produces a positive net thermal efficiency, but overall efficiency would probably decrease considerably (the original efficiency was 38.5%; for a 100 MW plant, the net electricity generated would be 81.8 MW, thus reducing the overall efficiency to [38.5 × 81.8%] = 31.5%) (Table 5.1). It is likely that optimization of the system would lead to a slight improvement in efficiency, but vast amounts of energy would inevitably be required for recovery and liquefaction.

Once the liquefied carbon dioxide has been forced to sink down to the deep ocean, it would continue to sink since it is heavier than the sea water under the tremendous pressures that exist at depths greater than 3 km. Even when the carbon dioxide is introduced at depths of about 500 meters, dissolution and diffusion occurs in the deep ocean and it is unlikely to return to the surface layers for 1,000 years due to global circulation patterns (for details, see Chapter 7). With regard to methods of introducing the carbon dioxide into the ocean, there has also been a report claiming that instead of carrying out liquefaction and then transporting the liquid via a pipeline, it would be better to transport liquefied carbon dioxide by ship to a point on the sea surface above the desired disposal site, and to then dispose of it there. This method has the advantage that disposal of the liquefied carbon dioxide is not restricted to just one site; on the other hand

Table 5.1 Energy evaluations for a 100 MW power plant using Steinberg's proposed process for recovery, liquefaction, and ocean disposal of carbon dioxide [*adapted from* Steinberg, 1984].

	Ocean disposal	
	at 500 m	at 3,000 m
Energy requirements for CO_2 control system (MW)		
Heat required for CO_2 recovery (electricity equivalent)	7.9	7.9
Electrical power required for CO_2 recovery	1.1	1.1
Electrical power required for liquefaction of CO_2	7.7	8.3
Pumping power required for transportation of CO_2	0.3	0.9
Total energy consumption	17.0	18.2
Net power plant capacity	83.0	81.8
Electricity production costs (mills/kWh in 1980 U.S. dollars)		
Power plant with no CO_2 recovery	32.0	32.0
CO_2 recovery	16.0	16.0
CO_2 liquefaction	7.3	7.6
Piping	1.5	1.8
Total capital investment	56.8	57.4

there is the worry that an increase in the number of sites of carbon dioxide injection would increase the acidic areas of the oceans. Ishitani *et al.* (1993) calculated that the later-mentioned recovery and isolation of carbon dioxide from the "integrated gasification combined cycle" would increase the cost of the generated electricity per kilowatt hour from 6.8 cents to 7.9 cents, and that the cost per tonne would be a little over $200 when including the costs of tankers and facilities on the ocean surface from which the carbon dioxide would be injected into the deep ocean by pipes.

Disposal in the form of dry ice (i.e. solid carbon dioxide) has also been investigated (Seifritz, 1988). Although the energy needed for producing the dry ice is double that for making liquefied carbon dioxide, dry ice is easy to handle; also, it has been calculated that if large lumps are used, by simply allowing the lumps to fall freely from the surface of the sea, most of the dry ice will reach the deep ocean in that state even though it will partially dissolve or undergo conversion into gaseous form. For example, if disposal took place into the fast-flowing Kuroshio current that flows off Japan, it has been calculated that the change in pH of the water would be kept to approximately 0.25, even if the total annual future amount of carbon dioxide emissions from Japan (estimated at 1.7 gigatons) were introduced in this fashion [currently emissions amount to 1.0 gigaton of carbon dioxide per year].

When this approach was originally put forward, the discussion was about storage of carbon dioxide in the form of a clathrate (a hydrate of composition $CO_2 \cdot 6H_2O$). This is because under deep ocean conditions a stable complex (i.e. the clathrate) is formed between water and carbon dioxide, in which the carbon dioxide is completely surrounded by water molecules. However, the current debate does not encompass storing all the carbon dioxide as clathrates. Instead, liquid carbon dioxide would be injected into the ocean at depths greater than 3,000 meters, from where it would sink to a greater depth or else accumulate on the sea bed; at this time a clathrate would be formed at the boundary between the liquid carbon dioxide and the sea water (i.e. the carbon dioxide would become coated with clathrates). There would of course be no problem for marine ecosystems as long as none of the carbon escaped into the sea water; also, it has been calculated that virtually no difficulties would occur even if the injected carbon dioxide were completely dispersed. Localized problems do occur, however, during the sinking of the liquid carbon dioxide and at the time of leakage from the accumulated carbon dioxide at the sea bed, and thus the formation of a clathrate would be a cause of some optimism. Furthermore, during surveys in the Mid-Okinawa Trough in 1989, the submersible survey vessel *Shinkai 1000* discovered that clathrates are formed naturally in the deep ocean (Sakai *et al.*, 1990).

The reaction described by Eq. (4) is also expected to lead to mitigation of the environmental effects, but since the reaction only takes place slowly, and as the deep waters of the Pacific Ocean have been below the surface for a long

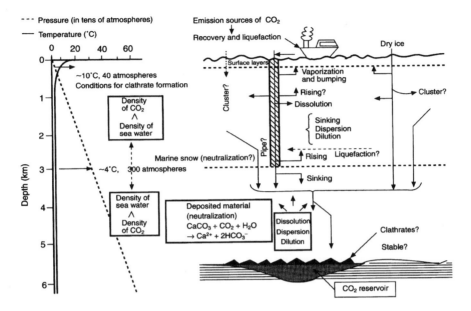

FIGURE 5.11 Ocean properties related to the ocean disposal of carbon dioxide [*adapted from* Nishio 1992].

time and are acidic, virtually no calcium carbonate exists at depths greater than 3,000 meters. Thus there are only faint hopes that this chemical reaction will prove to be of benefit.

In the future this technique needs to be further developed and evaluated, but no matter how the research progresses, the approach necessarily involves the use of the deep ocean, where human beings have neither lived nor even observed in detail, and also the expenditure of extra valuable energy resources (in order to carry out the recovery and injection procedures). In addition, after 1,000 years of storage, the carbon dioxide will in due course be returned to the atmosphere. This means that the technique can only be an emergency measure, and it is to be hoped that the day will never come when it needs to be put into practice.

The various physical phenomena associated with the ocean disposal of carbon dioxide are summarized in Fig. 5.11.

5.7 SUBTERRANEAN STORAGE

Besides the ocean, various other sites are under consideration for the disposal and storage of carbon dioxide, such as disused oil wells, abandoned salt mines, and depleted natural gas wells, but problems surround the reliability of both the method and transportation procedures. The advantage of such methods is that they alleviate the need for compression on the scale used in ocean disposal, therefore making them attractive from the energy point of view. However, in each case this raises the fear that carbon dioxide may be released on land. Even if the gas is temporarily dissolved in water under high pressure, it is possible that a release of the gas could result in a disaster due to the so-called champagne phenomenon, whereby the dissolved carbon dioxide gushes forth as gas follow-ing a fall in pressure, causing a shortage of oxygen. Furthermore, Table 5.1 shows that even in the case of subterranean storage, enormous amounts of energy are required for isolation of the carbon dioxide which, while not on the same scale as the amounts involved in injection into the deep ocean, mean that the approach must still be regarded as only an emergency measure.

The first method of storage is the secondary or tertiary recovery of oil known as enhanced oil recovery. The primary purpose of this technique is actually the increased production of oil, and not the isolation of carbon dioxide. In this method, the carbon dioxide is pressurized so as to mix with the oil; the viscosity of the oil thus decreases and moves more easily, and the oil is then forced out. Naturally some of the carbon dioxide is discharged along with the oil, but this can also be recycled. The subterranean pressure at which the majority of the pressurized carbon dioxide would be stored dictates that storage must be in oil fields at depths greater than several hundred meters, and that no leakage of the carbon dioxide occurs. Several of the carbon dioxide recovery techniques which will be discussed later were basically developed for use in such enhanced oil recovery.

Figure 5.12 shows the plan envisaged for recovering carbon dioxide during enhanced oil recovery. It has been calculated that on a worldwide scale it is pos-sible to treat nearly 63.0 gigatons of carbon dioxide (20.0 gigatons of carbon equivalent). However, the oil reserves in the oil fields which will be subject to

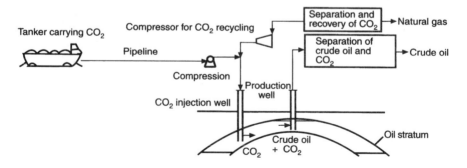

FIGURE 5.12 Use of carbon dioxide in enhanced oil recovery and subter-ranean sequestration [*adapted from* Tanaka, 1992].

this treatment amount to 2×10^{12} barrels (3×10^{14} liters), which means that there will be a future release of a further 200 gigatons of carbon equivalent in the form of carbon dioxide. While it may be desirable from the viewpoint of effective use of carbon dioxide, it is only possible to store one-third of the carbon dioxide released, which falls far short of the requirements. At the present time, much of the supply of carbon dioxide for use in enhanced oil recovery comes from the use of natural carbon dioxide layers; at this time, the supply of carbon dioxide is unlikely to result in greater costs, but any recovery and transportation of the carbon dioxide from the site of energy consumption would entail the extra costs shown in Table 5.1 and Fig. 4.3.

Another method of storage is the use of depleted gas wells. In the Netherlands, disused natural gas fields have been considered as a candidate for a disposal site, as this would facilitate the storage of 45 years' worth of domestic carbon dioxide production. The costs incurred in doing this are estimated to amount to only a small fraction of the total cost, when including the costs of carbon dioxide recovery; in other words, the recovery process involves extremely high costs. The use of natural gas pipelines has been proposed to address the problems of transportation.

Even at the present time, countries such as the United States are storing natural gas underground, with the amounts involved being approximately equal to half of Japan's annual production of carbon dioxide. On a global scale, it is estimated that it would be possible to store around 180 gigatons of carbon dioxide (50 gigatons of carbon equivalent). The advantage of this method is that use would be made of sites in which existing methane gas deposits have not experienced any leakage, which should mean that the sites would be extremely reliable. However, one mole of methane converts into one mole of carbon dioxide; thus, although there are naturally some pressure differences, the amount of carbon dioxide that can be stored is only about the same as the natural gas that is extracted, which means that it is impossible to additionally cope with the carbon dioxide released from coal etc.

The main constituents of natural gas are usually hydrocarbons such as methane, but in the actual operation of LNG (liquefied natural gas) plants, carbon dioxide frequently accounts for over 10% of the gas mixture used as a raw material. This author has heard of a planned Indonesian plant in which over half of the gas will be carbon dioxide. Such carbon dioxide at least should be returned either to the original gas field or else to aquifers. (An aquifer is a layer of rock or soil which is able to hold or transfer large quantities of water.) Indeed, some of the plans for this type of LNG project do indeed incorporate such measures.

A method for utilizing aquifers which was proposed by some Japanese researchers (Koide *et al.*, 1992) has recently been attracting attention. Aquifers are formed when sea water etc. becomes trapped within rock formations; in many cases, they exist in the vicinity of oil and gas fields, and often contain organic materials which originate from water-soluble natural gas (the main constituent of which is methane). Even in Japan, some of the water-soluble natural gas is used commercially; the reserves are estimated to be equivalent to about 800 billion cubic meters under normal conditions of temperature and pressure. Under pressure, carbon dioxide is much more soluble than methane; Koide reported that if the methane is extracted and replaced by carbon dioxide under pressure, it is possible to dissolve 15 times as much carbon dioxide as methane. This would correspond to 24.0 gigatons of carbon dioxide, which would be equal to between 20 and 30 years of emissions in Japan at current levels. Just by using a mere 1% of the sedimentary basins worldwide, it would be possible to store 320 gigatons of carbon dioxide, equal to over 10 years' worth of global emissions. During storage, the rock would undergo weathering reactions, and thus it is likely that the acidity caused by the carbon dioxide would be neutralized. Calculations show that the electricity needed for such treatment (i.e. not counting the energy required for separation and recovery) would come to 5.8% of the amount generated by coal-fired plants at a cost of 2.1 cents per kilowatt hour, but the majority of electrical power could be provided by the natural gas which would be obtained (Koide, 1992). Before implementing such schemes, it would be necessary to investigate the potential environmental effects from various angles, particularly the problem of leakage (e.g. caused by possible weakening of the rock formation due to the above-mentioned weathering). However, the approach can be considered a possible method that would incorporate the use of unused resources (e.g. methane which is dissolved in water).

Figure 5.13 shows a system for subterranean storage. Table 5.2 gives a summary of the above, and shows the total amounts available for disposal and storage, as well as the potential annual amounts that can be re-utilized (it should, however, be pointed out that there are slight variations in the estimates put forward by different researchers).

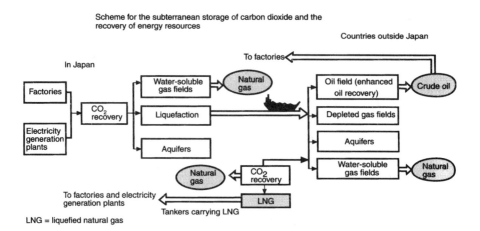

FIGURE 5.13 Subterranean storage of carbon dioxide [Koide, 1992].

Table 5.2 Carbon dioxide disposal and utilization options [International Energy Agency, 1993].

Disposal method	Potential storage (GtC)	Carbon dioxide use	Quantity (GtC/yr)
Oceans	2×10^7	Enhanced oil recovery	0.1
Aquifers	87.0	Biological processes	0.5
Exhausted oil and gas wells	29.0	Chemicals	0.05

GtC = gigatons of carbon.

5.8   RECOVERY OF CARBON DIOXIDE FROM FLUE GASES

The following is a summary of the methods of recovering carbon dioxide from flue gases from the viewpoints of cost and energy. Besides the monoethanolamine (MEA) absorption process which was used in Steinberg's calculations (1984), various other methods have also been proposed e.g. the use of other absorbents, physical absorption, cryogenic separation, membrane separation, pressure swing adsorption (PSA), and temperature swing adsorption (TSA). There are problems concerning both the need to reduce costs, and also the need for pre-treatment to remove sulfur oxides, soot and dust when coal is combusted; in the majority of cases, however, this merely involves further development of existing technology, and unfortunately there does not seem to be any hope of substantial improvements in efficiency. On the other hand, most of these techniques were developed for use in enhanced oil recovery, or in order to produce carbon dioxide of almost 100% purity for purposes such as dry ice production or for food processing; consequently, it is necessary to adequately investigate whether the carbon dioxide concentrations involved and the existence of impurities will restrict the application of the techniques. The special features of the different methods, and a comparison with the oxygen combustion method (which is discussed in the next section), are listed in Table 5.3; the estimated costs and efficiency of electricity generation are given in Table 5.4.

This author's own opinion is that there is not enough justification for pursuing research into these techniques on an urgent basis. First, the carbon dioxide, which is present in natural gas and must be separated from it, is directly obtained in gaseous form with a purity of almost 100%, after which it is stored. After ensuring that this is carried out in a safe manner, it is then necessary to develop the most appropriate means of storage. In any event, at the moment the techniques for recovery, disposal and storage can only be regarded either as emergency measures, or else as techniques which may be regretted in the future

Table 5.3 Summary of isolation and recovery techniques for carbon dioxide [adapted and expanded from Nakayama *et al.*, 1993].

---

1. Alkanolamine process (chemical absorption from gases emitted from fossil fuels)
   a. Key points of process
      —Absorption of CO_2 by a weakly alkaline solution of alkanolamine at $40-50°C$ under atmospheric pressure
      —Regeneration of CO_2 and the absorbent liquid by heating at $100-120°C$
   b. Advantages and disadvantages
      —Good for absorption at atmospheric pressure (technique is applicable over a wide range of pressures)
      —Large-scale operations are comparatively simple
      —Care must be exercised when SO_x is present in the exhaust gases since sulfur compounds cause degradation of the absorbent
      —Highly corrosive
      —Technique widely applied e.g. at natural gas plants and oil refineries
   c. Typical processes etc.
      —Monoethanolamine (MEA) process
      —Amine Guard process

2. Heated potassium carbonate process (chemical absorption from gases emitted from fossil fuels)
   a. Key points of process
      —Absorption of CO_2 by potassium carbonate solution at $70-120°C$ under pressure
      —Regeneration of the absorbent liquid by heating under reduced pressure
   b. Advantages and disadvantages
      —Good for absorption at high pressures
      —Large-scale operations are comparatively simple
      —Requires less heat than the amine process
      —Sulfur compounds do not cause much degradation of the absorbent
      —Highly corrosive
      —Technique widely applied e.g. at natural gas plants and oil refineries
   c. Typical processes etc.
      —Benfield process
      —Catacarb CO_2 process

3. Physical absorption from gases emitted from fossil fuels
   a. Key points of process
      —Physical absorption of CO_2 by solutions of methanol or polyethylene glycol etc. at high pressures and low temperatures (i.e. lower than room temperatures—e.g. at approx. $5°C$)
      —Regeneration of the absorbent liquid by heating under reduced pressure
   b. Advantages and disadvantages
      —Good for absorption at high pressures and low temperatures
      —Large-scale operations are comparatively simple
      —Requires less heat than the amine process
      —Sulfur compounds do not cause much degradation of the absorbent
      —Only slightly corrosive

Table 5.3 (*continued*)

c. Typical processes etc.
—Rectisol process
—Sepasolv MPE process (BASF, Germany)
—Selexol process

4. Adsorption from gases emitted from fossil fuels
   a. Key points of process
      —Adsorption of CO_2 by zeolite and carbon-based adsorbents
   b. Advantages and disadvantages
      —Simple operation and maintenance
      —Few plants currently employ this technique on a large scale
      —Two or more steps are required due to the low recovery rate
      —Pre-treatment is required to remove SO_x and water vapor etc.
   c. Typical processes etc.
      —Pressure swing adsorption (PSA) process
      —Temperature swing adsorption (TSA) process
      —Pressure and temperature swing adsorption (PTSA) process (this is a combination of the PSA and TSA processes)

5. Membrane separation from gases emitted from fossil fuels
   a. Key points of process
      —Selective separation of CO_2 using a membrane composed of polyamide or cellulose acetate etc.
      —Another technique involves the use of a liquid membrane
   b. Advantages and disadvantages
      —Simple process
      —Few plants currently employ this technique on a large scale
      —Requires high pressures (17–35 atmospheres)
      —Two or more steps are required due to the low recovery rate
      —Problems exist with the durability of membrane materials
   c. Typical processes etc.
      —Under test by British Coal and other companies

6. Combustion under oxygen diluted with CO_2 (direct recovery from exhaust gases)
   a. Key points of process
      —Recovery of CO_2 by combusting coal under oxygen to produce a CO_2 concentration in excess of 95%, and then recovering that gas mixture
      —The oxygen concentration for combustion is adjusted by adding recycled exhaust gas to pure oxygen
   b. Advantages and disadvantages
      —No consideration needed regarding the effect of impurities such as SO_x etc. in the exhaust gases
      —Suitable for fossil fuel combustion processes
      —Comparatively simple to adapt to large-scale operations
      —Thorough investigation is necessary since this is a new system
      —Still at the stage of basic research
   c. Typical processes etc.
      —Under test in Japan at the Center for Coal Utilization and at other institutions

Table 5.4 Carbon dioxide capture study summary for the pulverized fuel plus flue gas desulfurization study [adapted from International Energy Agency, 1993].

Separation technique	Efficiency (%)	Power cost* (mills/kWh)	Cost of CO_2 avoided** ($/tonne)	Cost of CO_2 recovered ($/tonne)	Recovery of CO_2 (%)	Emission rate of CO_2 (gCO_2/kWh)
PF + FGD base case***	39.9	50	N/A	N/A	0	829
+ membranes	31.1	80	50	40	80	194
+ membranes and MEA	29.7	70	40	30	80	222
+ MEA ****	29.1	80	50	30	80	227
+ PSA	28.5	110	80	60	95	57
+ TSA	29.5	180	260	170	70	335

Note: *1 mill = one-thousandth of $1.
**Includes the cost of recovery of that CO_2 produced when generating the energy needed for the CO_2 recovery process.
***(pulverized coal-fired power generation + desulfurization).
****Data for cryogenics was not applicable.
PF = pulverized fuel, FGD = flue gas desulfurization, MEA = monoethanolamine, PSA = pressure swing adsorption, TSA = temperature swing adsorption.

(using the term 'regret' in the sense described in Section 2.7). The area which most warrants investigation concerns the question of where storage should take place. (For a more detailed discussion of the above recovery techniques, please refer to sources listed in the bibliography.)

5.9 NEW FOSSIL FUEL-BASED PROCESSES OF ENERGY PRODUCTION WHICH INCORPORATE THE RECOVERY OF CARBON DIOXIDE

At present, combustion is performed in air; a technique which is attracting attention in the event that carbon dioxide recovery becomes a requirement is combustion under an atmosphere of oxygen (Fig. 5.14). The combustion of fossil fuels under an oxygen atmosphere results in excessively high temperatures; in order to lower the temperature of combustion it is necessary to dilute the oxygen by recycling the carbon dioxide. This system is noteworthy for the following reasons. First, apart from water, virtually the only product of the process is carbon dioxide. Therefore if this carbon dioxide is simply to be compressed and stored in a gas field, there is absolutely no need to perform separation procedures for carbon dioxide or for toxic materials such as SO_x since these will not be released into the atmosphere. However, if, for example, the carbon dioxide were to be liquefied for ocean disposal, the water content would have to be removed prior to liquefaction in order to prevent the compression of the carbon dioxide becoming difficult.

Next, let us consider the situation in which pollutants are removed from the gas in order to avoid contamination of the ocean. Since only oxygen is used in the combustion process, and since thermal NO_x (nitrogen oxides) are only formed from the nitrogen in the atmosphere, no thermal NO_x is produced as a result of the oxygen combustion. The only NO_x which is formed will originate from the nitrogen in the fossil fuels (fuel NO_x); this will lead to a considerable

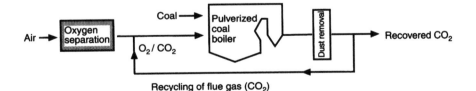

FIGURE 5.14 Flow diagram for the recovery of carbon dioxide based on pulverized coal combustion under an oxygen-rich atmosphere [Shirakawa and Noguchi, 1992, *adapted from* Wolsky, 1986].

decrease in the total amount of NO_x present, thus making the removal of NO_x simple. Desulfurization is also greatly simplified since it can be carried out at the same time as liquefaction. On the other hand, however, a great deal of energy is needed to extract the oxygen from the air in order to carry out this process. The estimated efficiencies of this and other processes are compared in Table 5.5. Although a considerable reduction in efficiency is unavoidable, the efficiency is still greater than that for the process of separating carbon dioxide from the flue gases which is considered in Table 5.4. However, it is impossible to arrive at a firm conclusion since the estimations of efficiency and cost vary greatly with the conditions selected for analysis.

A research group in the Netherlands (Block *et al.*, 1989) has proposed a process for recovering carbon dioxide using an integrated gasification combined cycle (IGCC). In this process, a stream of oxygen is first blown into a coal gasifier; the gas mixture that emerges contains carbon monoxide, 90% of which is converted into carbon dioxide by means of a two-step shift reaction. (Here, 'shift', or perhaps more correctly 'water shift' or 'carbon monoxide shift', refers to the shift from left to right in the reaction $H_2O + CO = CO_2 + H_2$.)

After this reaction takes place, the carbon dioxide is separated from hydrogen. In the process, oxygen is used for the gasification, but since the coal is not completely combusted, the energy required for oxygen production is less than in the previously described combustion under oxygen. Furthermore, the extra efficiency obtained by the adoption of a combined cycle more than compensates for this energy. Indeed, gasification with oxygen is common (even when carbon dioxide recovery is not performed) because the energy required to compress oxygen is less than that needed for compression of air. On the other hand, the gases which

Table 5.5 Numerical results from base case power generation studies for a standard station size of $500\,MW_{SO}$ [International Energy Agency, 1993].

Study	Fuel	Efficiency (%)	CO_2 (g/kWh)	CO_2 (vol%)	Specific cost (US$/kW)
PF/FGD	Coal	39.9	830	13	900–1,200
CCGT	Gas	52.0	410	3	600–800
IGCC	Coal	41.7	790	15	1,600–1,900
O_2/CO_2	Coal	32.8	1,010	90	2,200–2,500

PF = pulverized fuel.
FGD = flue gas desulfurization.
CCGT = combined cycle gas turbines.
IGCC = integrated gasification combined cycle.
O_2/CO_2 = combustion in an atmosphere of recycled CO_2 and oxygen.

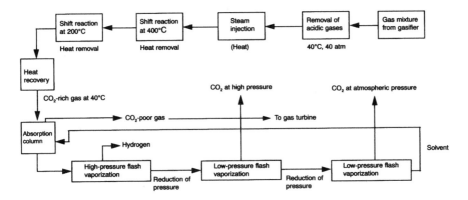

FIGURE 5.15 Integrated gasification combined cycle with subsequent recovery of carbon dioxide [Block *et al*., 1989].

are given off during the combustion of fossil fuels include nitrogen; but since nitrogen is not given off when the IGCC process is carried out using oxygen, the resultant gas has an increased concentration of carbon dioxide, making it easy to separate the carbon dioxide. It is thus possible to apply the Selexol process, which is one of the previously-mentioned methods of physical absorption (Selexol is the trade name for the dimethyl ether of polyethylene glycol). Alternatively, it is possible to separate hydrogen from carbon dioxide by means of a membrane. That is, with regard to the recovery of carbon dioxide, the IGCC method has an advantage over techniques that employ either separation from waste gases or combustion under oxygen. A flow diagram of the proposed IGCC process with carbon dioxide recovery is shown in Fig. 5.15.

When devising systems for recovering carbon dioxide, it is necessary to modify the processes in order to maximize the overall efficiency, including the processes for converting fossil fuels such as coal. This re-designing of systems, together with the previously discussed problems of where to perform isolation and how to carry out fixation, are extremely important topics that require further investigation.

5.10 A SYSTEM THAT CENTRALIZES EMISSION SOURCES AND INCORPORATES THE RECOVERY OF CARBON DIOXIDE

This chapter has considered various processes for the recovery and isolation of carbon dioxide. However, with the exception of separation procedures for limited purposes (e.g. the use of carbon dioxide for enhanced oil recovery), these processes have not been put into practice. At this point it should perhaps be noted that although the discussion so far has considered carbon dioxide recovery at the sources of large-scale emissions such as thermal power stations, a greater percentage of carbon dioxide emissions result from household heating, cooking, and transportation (i.e. automobiles). In practice, it is impossible to recover carbon dioxide directly from these sources of emissions. However, if it were possible to recover carbon dioxide from the sources of large-scale emissions and then store this (in the oceans etc.), then implementation of the following system would make it possible to curb carbon dioxide emissions from small-scale sources such as homes and automobiles. This is what is meant by centralization of emission sources.

As was stated in Chapter 3, hydrogen is not primary energy, but carbon is always present in fossil fuels. Thus let us consider, for example, if instead of the current use of town gas, hydrogen were produced from fossil fuels and water at the site of carbon dioxide emission (e.g. chemical plants). While the water consumed the energy and formed hydrogen, the carbon would be converted into carbon dioxide. This carbon dioxide would be recovered during the process itself and then isolated. However, in addition to the previously-mentioned problem of the final site of carbon dioxide storage, there are many other problems that would need to be resolved, such as those relating to the safety aspects of employing hydrogen for household use and possible leakage in supply lines).

The main fuels currently used for transportation are gasoline and light oils. In future, it is also possible that coal oil (i.e. the oil produced by coal liquefaction) might be used, but no matter what the precise type of oil is used, it would lead to the production of carbon dioxide. If hydrogen is used for vehicles instead of gasoline or light oils, it will be necessary to develop appropriate engines and adequate safety measures. With respect to not only hydrogen-powered cars but also electric cars, the implementation of carbon dioxide recovery systems at the time of electricity generation would make it possible to achieve virtually complete elimination of carbon dioxide emissions. Furthermore, even from the viewpoint of overall energy efficiency, electric cars would be superior to current gasoline-powered cars. Another possible intermediate measure would be to reduce carbon dioxide emissions by using low-carbon liquid fuels such as methanol.

Finally, it should be emphasized that all of these methods have been discussed on the premise that carbon dioxide is recovered and isolated. However, each of the various secondary energy systems was investigated before the carbon dioxide issue materialized, and none of them have been put into operation on a commercial basis. It is possible that the emergence of the carbon dioxide problem will overturn the foundations of earlier evaluations. However, it is important to distinguish between construction of a secondary energy system that aims at increasing the overall energy efficiency (i.e. a system that will not be regretted later), and a secondary energy system that forms part of a system which, though including a centralization of emission sources, will indeed lead to later regret (although this might be acceptable as an emergency measure).

This whole problem is an environmental issue of global proportions. It cannot be resolved merely by curtailing emissions from individual automobiles if these are then replaced by emissions from electricity generating plants. It is not simply a pollution problem, but a question of overall energy use.

References

Arakawa, Y. 1990. In Komiyama, H. *et al.* (eds.). 1990. A Handbook of the Global Warming Issue, p. 361, IPC, Tokyo, Japan (in Japanese)

Block, K., C. A. Hendricks and W. C. Turkenburg. 1989. Proceedings of Energy Technologies for Reducing Emissions of Greenhouse Gases (seminar), Paris, April 1989

Fujishima, A. 1993. *Proceedings of The 8th National Congress for Environmental Studies.* p.111 (in Japanese)

Haugan, P. M. and H. Drange. 1992. *Nature,* **357**, 318. Cited in Ohsumi, T. 1993. *Kagaku* **63**(1), 17 (in Japanese)

International Energy Agency (IEA). 1993. *Greenhouse Issues, No. 7.* IEA ISSN 0967-2701, March 1993

International Energy Agency (IEA). 1996. *Greenhouse Gases: Mitigation Options Conference.* London, August 1995. *Energy Conversion Management* **37**(6–8), 1996

Ishitani, H., R. Matsuhashi, A. Ohmura and K. Takeda. 1993. *Energy and Resources* **14**, 85 (in Japanese)

Koide, H. 1992. *Chemical Engineering* **37**(10), 56 (in Japanese)

Koide, H., Y. Tazaki, Y. Noguchi, S. Nakayama, M. Iijima, K. Ito and Y. Shindo. 1992. *Energy Conversion Management* **33**, 619

Kojima, T. 1995. *Energy Conversion Management* **36**, 881

Kojima, T., N. Nagamine, N. Ueno and S. Uemiya. 1996. *Energy Conversion Management* (In press)

Nakayama, S., S. Miyamae, U. Maeda and T. Tanaka. 1993. *Energy and Resources* **14**, 87 (in Japanese)

Nishio, M. 1992. *Chemical Engineering* **37**(10), 50 (in Japanese)

Ohsumi, T. 1991. *Journal of the Fuel Society* **70**, 225 (in Japanese)

Sakai, H., T. Gamo, E.S. Kim *et al.*, 1990. *Science* **248**, 1093

Seifritz, W. 1988. *Proceedings of World Hydrogen Energy Conference*, p. 1497. Cited in Akai, M. 1990. In Komiyama, H. *et al.* (eds.). 1990. A Handbook of the Global Warming Issue, p. 433, IPC Pubs., Tokyo, Japan (in Japanese)

Shirakawa, H. and Y. Noguchi. 1992. Preprints of The Second Meeting of Coal Utilization Technology, p. 235. Sep. 3rd–4th, 1992, Tokyo, Japan (in Japanese)

Steinberg, M. 1984. A system study for the removal, recovery and disposal of carbon dioxide from fossil fuel power plant in the U.S. United States Department of Energy (DOE) Report DOE/CH/00016-2. Cited in Komiyama, H. *et al.* (eds.). 1990. A Handbook of the Global Warming Issue, p. 343, 433, IPC, Tokyo, Japan (in Japanese)

Takeuchi, H. 1992. *PPM* (10) 34 (in Japanese)

Tanaka, S. 1992. *Energy Conversion Management* **33**, 587

Tanaka, T. 1989. *Geology News* **4**(22), 60 (in Japanese)

Tanaka, T. 1993. Presentation at Japan–China International Symposium on Study of Mechanism of Desertification, Tsukuba, Japan, March 2–4, 1993

Wolsky A.M. 1986. A New Method of CO_2 Recovery. In *Proceedings of the 79th Annual Meeting of the Air Pollution Control Association*, Minneapolis, U.S.A., June 22nd–27th 1986

6 THE ENERGY PROBLEM AND CARBON FIXATION BY LAND VEGETATION

This chapter will review the technologies which the author hopes or believes will prove the most beneficial.

6.1 EFFECTIVENESS OF CONVERSION OF SOLAR ENERGY BY LAND VEGETATION

First let us consider the forest ecosystem, which plays the greatest role in the carbon dioxide balance of any vegetation. Forests are important because of photosynthesis, which can be considered to be a process that involves the conversion of solar energy, with a re-circulation of materials and an energy flow (Fig. 6.1). The figure considers the case of tropical rainforests, which are believed to have the greatest potential for carbon dioxide fixation. Tropical rainforests are drenched with sunshine, and during the daytime approximately 1% of this is used for photosynthesis. About 75% of the products of photosynthesis are consumed on the spot during the same day for respiration. The remainder is either stored in the leaves and branches or used for growth, but in the space of just one year approximately 20% of the products of photosynthesis descend into the earth and undergo decomposition. The rest (i.e. around 5% of that originally produced by

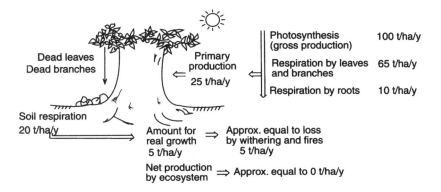

FIGURE 6.1 The cycle of materials in forests (based on the quantity of dry organic matter) [Oikawa, 1989].

145

photosynthesis) is used for the growth of the trunk and branches. The total amounts to only 0.05% of the solar energy received; the amount that human beings can usually utilize is just a fraction of this. Even if it is supposed that all of this (including even the fallen leaves) can be used as biomass, it is still only possible to use 0.25% of solar radiation. The efficiency of forests as a place of energy production is thus remarkably low in comparison with the efficiency of photovoltaic cells, which at present is easily in excess of 10% and which may even exceed 20% in the future. That is, in terms of unit land area there is simply no contest. (While trees are young, growth certainly takes place a little more rapidly, and more energy is accumulated; however, the amount is unlikely to be large.)

One more important fact is that in mature natural forests the accumulated carbon returns to the earth after the trees die because of fires, volcanic activity, changes in the earth's crust, old age, or disease etc. Consequently the overall production of organic matter (i.e. the net energy accumulation) is zero. In actual fact, even if this were not zero, the rate of net production of organic matter (due to photosynthesis) might be one millionth the rate at which energy is used in the form of fossil fuels, given that humans are likely to use up in a few hundred years the enormous amount of coal which has gradually accumulated from trees on the earth's surface over a period of several hundred million years. Considering that this calculation includes the coal age (which due to its high carbon dioxide concentrations and high temperatures witnessed an accelerated rate of carbon accumulation), the present rate of accumulation of carbon or energy by forests might be less than one millionth of the rate of consumption of energy (i.e. the rate of conversion of carbon into carbon dioxide).

6.2 THE IMPORTANCE OF RAINFORESTS

If the above were summarized in a slightly provocative manner, and the argument moreover extended to include points that will be dealt with later, the following conclusions would be reached:

(1) Land vegetation contains three times as much carbon as the atmosphere. Furthermore, ten times the amount of carbon dioxide released annually from fossil fuels is exchanged with the atmosphere by means of photosynthesis. However, in the final analysis this is used in the course of everyday life; it is extremely difficult to store this energy as organic matter.

(2) The net rate of fixation of carbon dioxide is zero in forests which are not utilized by humans. At current rates of fossil fuel consumption, coal cannot accumulate until the passage of time equal to a full one million times the period during which it is being consumed.

(3) The Amazon forest will always remain the lungs of the Amazonian ecosystem, but not the lungs of the entire earth. Excluding the release of carbon dioxide from fossil fuels, it is the areas of agricultural land that are the lungs for the activities of human beings.

(4) The efficiency of solar energy use in mature tropical rainforests is at most only 0.25%, even when all the net primary production of organic matter is considered. Only about 20% of this is retained in the tree in the form of branches and trunks. Accordingly, the forests do not have a large role to play when considering vegetation as a base for the conversion of solar energy.

(5) The total energy consumption by human beings is one-fifth of the net primary production by all vegetation in the world. That is, if all the branches and trunks from all the world's vegetation were used, it would in the end equal the total amount of energy used by human beings.

(6) The productivity of the tropical rainforests is one of the highest anywhere in the world. Consequently, it generally makes no sense to consider using other areas for the production of biomass for use as a source of energy.

From this, it might be concluded that even if agricultural land is necessary for the production of food, forests are unnecessary. But is such a conclusion correct? Let us reconsider the conclusion of Chapter 2. The destruction of tropical rainforests means the release of 1.5–2.0 gigatons of carbon equivalent of carbon dioxide. That is equal to half the 3.5 gigatons accumulated in the atmosphere each year. Thus prevention of the destruction of the tropical rainforests would lead to a 50% reduction in the amount of carbon dioxide released into the atmosphere each year. Furthermore, if afforestation could proceed at the same rate as the destruction of existing tropical rainforests, this would resolve the carbon dioxide problem. This therefore means that afforestation is extremely important in the carbon dioxide problem.

We therefore have to examine why we have these two inconsistent conclusions.

6.3 THE CARBON BALANCE, AND THE ROLE OF VEGETATION

Let us restate the role of land vegetation such as forests in the carbon balance and carbon cycle, which consist of the stock of ecosystems (including the living land vegetation and the soil) and the flow (the exchanges with the atmosphere involving photosynthesis and respiration—or combustion and decomposition). The photosynthesized organic matter which is not immediately used but which is accumulated in the form of leaves etc. is referred to as the net primary production. Leaving aside the question of the exact amount of time involved, the net primary production remains a part of the stock for a certain period of time. The stock exists in the form of living organisms and dead bodies contained within the earth, both of which are composed of organic matter. Furthermore, it would appear correct to count fossil fuels such as coal as part of this stock since these make up the final form of fixation. However, this is different from the living organisms and dead bodies contained within the earth in that exchange with the atmosphere is virtually zero during the time scale under discussion. Within the range of the present discussion, there is an identical amount of flow in both directions under steady state conditions (i.e. in natural forests), and the total stock remains unchanged. This is a point additional to the factors indicated in Fig. 6.1. Taking a broad view of the situation, which includes a consideration of human beings receiving the abundant products of the forests etc., human beings are also part of the same ecosystem.

However, at the present time the importance of forests is under discussion because a steady state no longer exists. There should naturally be almost no exchange between the atmosphere and the carbon that has undergone fixation as a fossil fuel, but the latter has indeed been gradually depleted, and the stock of living and dead bodies which are continually undergoing exchange with the atmosphere is gradually decreasing.

Table 6.1 classifies each of the surface ecosystems in the world (including the oceans). It gives the sizes of the areas they cover, the carbon-equivalent stock of living vegetative matter and dead substances, the density (the total stock divided by the area), the net primary production (the flow, in terms of gigatons of carbon equivalent per annum), and the amount of production per unit area (the production density). Furthermore, the time constant (i.e. the minimum time required for a complete exchange of carbon stock) is obtained by dividing the stock by the flow; the value is given both for living matter, and for living matter plus dead matter combined. Also, in cases where two values are presented, the first figure is that reported by Yoda (1982), with the figure in parentheses at the side or below being data for living matter reported by Woodwell (1987) or for dead matter reported by Tsutsumi (1987).

From differences in the data quoted from Woodwell for the amount of carbon in living matter in tropical forests and in tropical grasslands (savanna), it was calculated that conversion of the tropical forests into grasslands would result in the loss of 17.0 gigatons of carbon per hectare (details given in Chapter 2). However, there are various estimates for this, and it is easy to see that the estimates are not very accurate.

Let us now consider the main points from Table 6.1. Even though the forested area amounts to only 34% of the total land surface (38% according to Woodwell), which represents only about 10% of the world's surface, its stock of living matter is extremely large, accounting for over 90% of all the world's living stock on land. In contrast, the amount of living stock in the world's oceans is small, and the living stock in the forests thus effectively accounts for 90% of the world total. The living stock in the tropical rainforests amounts to nearly 50% of the total for forests (and exceeds 50% according to Woodwell's data). If dead matter is also included in the calculations, the contribution of forests is 62% to the land total, and 42% to the total for the whole world. The percentage that forests account for in the flow equation is also naturally large, amounting to just under 70% for all land areas and a little over 40% for the whole world. The areas besides forests which have a high density when dead matter is included are swamps, tundra and temperate grasslands etc. On the other hand, forests account for the greatest density of primary production, followed by estuaries, which experience an influx of nutrient salts. Swamps are comparable to this, but in the table they have been listed together with freshwater ecosystems and on average are comparable to boreal forests (boreal forests are the coniferous forests that exist in the low Arctic regions). Here, agricultural land is unique in that even though the productivity (the flow) is comparable to boreal forests, the living stock per unit area amounts to just a small fraction, and even when dead matter is included is only 25–30% of the total for boreal forests. In addition, even though the living stock in the ocean is extremely small, the value of the flow is over half that of the land.

The time constant given when the total stock (shown in the final column of Table 6.1) is divided by the flow provides a necessary yardstick of time regarding the formation of that ecosystem. However, the rate of decomposition of dead matter in the earth varies greatly since chemical composition varies with depth, and a minute fraction of this is converted extremely slowly, first into peat, and then into coal. That is, the time constant must be regarded only as an indicator. However, with regard to living vegetative matter, this value can be taken to be a relatively good guide.

In comparison to aquatic environments, the time constant for land environments (and forests in particular) shows that over extremely long periods of time large quantities of carbon are accumulated. Also, whereas almost all of the organic materials are in the form of dead matter (i.e. organic matter in the process of undergoing decomposition in the oceans), over one-third of all the organic materials on land consist of living matter.

Thus when consideration is given to the organic matter on the land surface, it is impossible to neglect the amount of dead organic matter contained in the soil in forests etc. In boreal and temperate forests the quantity of organic matter in the soil (i.e. dead bodies) is comparable to, or even greater than, the amount of vegetative matter (i.e. living bodies), but in tropical rainforests the amount of organic matter is small (especially in the surface layers) and, as shown in Fig. 6.2, the rate of decomposition is high. That is, in cold regions the forest ecosystems are stabilized by the soil, but in tropical forests it is quite likely that loss of the surface would lead immediately to destruction of the ecosystem.

Reiterating the points made in Chapter 2, the average annual rate of loss of true tropical forest during the period 1981–1985 was 10 million hectares; if this is multiplied by the difference in surface carbon content between tropical forests and savanna (i.e. 170 tonnes per hectare) the rate of carbon released is found to

Table 6.1 Amounts of carbon stored in various ecosystems, the rate of net primary production, and time constants (TC).

Ecosystem	Area (10⁸ ha)	Organic matter (GtC)		Density (tC/ha)			Net production (GtC/yr)	Production density (tC/ha/yr)	TC* (yr)	
		Living	Dead	Living	Dead	Total			Living	Total
Tropical forests	18 [24.5]	270 [461]	126 [147]	150 [188]	70 [60]	220 [248]	13.6 [22.2]	7.5 [9.1]	20 [21]	29 [27]
Temperate forests	12 [12.0]	130 [174]	153 [108]	110 [145]	130 [90]	240 [235]	7.1 [6.7]	5.9 [5.6]	19 [26]	40 [42]
Boreal forests	13 [12.0]	110 [108]	225 [156]	85 [90]	175 [130]	260 [220]	4.3 [4.3]	3.3 [3.6]	26 [25]	79 [61]
Woodland and shrublands	8 [8.5]	40 [22]	80	50 [26]	100	150	2.4 [2.7]	3.0 [3.2]	17 [8]	50
Freshwater and swamps	4 [4]	4 [13.5]	80	10 [34]	200	210	1.25 [3.1]	3.1 [7.8]	3.2 [4]	67
Tropical grasslands	13 [15]	7 [27]	104	5 [18]	80	85	1.95 [6.1]	1.5 [4.1]	3.6 [4]	57
Temperate grasslands	9 [9]	9 [6.3]	135	10 [7]	150	160	2.25 [2.4]	2.5 [2.7]	4.0 [3]	64
Agricultural land	14 [14]	14 [6.3]	84	10 [4.5]	60	70	4.2 [4.1]	3.0 [2.9]	3.3 [2]	23
Tundra	8 [8]	4 [2.3]	160	5 [2.9]	200	205	0.4 [0.5]	0.5 [0.6]	10 [5]	410
(Semi) deserts	45 [42]	5 [6.1]	26	1 [1.5]	6	7	0.9 [0.7]	0.2 [0.2]	5.5 [9]	35
Abandoned land	5 [—]	15 [—]	40	30 [—]	80	110	1.25 [—]	2.5 [—]	12 [—]	44
All land areas	149 [149]	608 [826.5]	1,213 [—]	41 [55.5]	81 [—]	122 [—]	39.6 [52.8]	2.7 [3.54]	15.4 [15.7]	45 [—]

Table 6.1 *(Continued)*

Ecosystem	Area (10^8 ha)	Organic matter (GtC)		Density (tC/ha)			Net production (GtC/yr)	Production density (tC/ha/yr)	TC* (yr)	
		Living	Dead	Living	Dead	Total			Living	Total
Estuaries	1.4	0.63		4.5			1.0	7.1	0.63	—
Algal beds and coral reefs	2	0.54		2.7			0.7	3.5	0.77	—
Region of upwelling	0.4	0.004		0.1			0.1	2.5	0.04	—
Continental shelf	27	0.12		0.045			4.3	1.6	0.03	—
Oceans	332	0.45		0.014			18.7	0.56	0.02	—
All salt-water environments	361	1.74	900	0.048	25	25	24.8	0.69	0.07	36
All regions	510 [510]	610 [828]	2,100 [—]	12 [16]	41 [—]	53 [—]	64.4 [77.6]	1.26 [1.52]	9.5 [10.7]	42 [—]

Note: [ ] indicates data from a different source. Data compiled from [Yoda, 1982], [Whittaker and Likens *et al.*, 1973], [Woodwell *et al.*, 1978], [Woodwell, 1987], [Tsutsumi, 1989], etc.

*Time constant = amount of carbon/net production.

GtC = gigatons of carbon.

tC = tonnes of carbon.

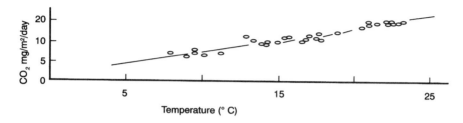

FIGURE 6.2 Changes with temperature in the rate of decomposition of organic matter in soil [Tsutsumi, 1987].

be 1.7 gigatons per year. On the other hand, a comparison based on the carbon amounts shown in Table 6.1 between the total amount of carbon (including dead matter) in tropical forests and that in deserts reveals a difference of 21 gigatons. This means that if the destruction of the forests proceeds all the way to desertification, it is possible that the above figure for the rate of carbon release will increase still further.

To summarize, the role of vegetation in the carbon dioxide problem is that of a storehouse. Accordingly, if the storehouse were to be destroyed the contents within it would be released to the atmosphere. If the storehouse were then to be constructed again, the contents would have to be absorbed from the atmosphere. Even at the present time the amount of carbon dioxide released by deforestation is equal to approximately one-third (or more) of the amount released from fossil fuels. In terms of cumulative amounts, the quantity released in the past due to deforestation was over double that from fossil fuels. Even to just return to the previous situation would mean an enormous accumulation of carbon.

Let us consider one further piece of evidence that supports this argument. Amidst all the clamor about the degree of destruction of tropical forests, it must be remembered that (as explained in Chapter 2) one theory holds that plant life actually absorbs carbon dioxide. It states that the missing carbon is absorbed by vegetation (which then becomes larger) or else is accumulated within the soil. In reality, though, in the context of the carbon dioxide problem this is extremely valuable evidence that the contribution of plant life is not to the flow but to the stock.

6.4 AREAS AVAILABLE FOR AFFORESTATION AND EFFECTIVE PERIODS FOR COUNTERMEASURES

Some of the opponents of countermeasures that utilize afforestation argue that forests absorb carbon dioxide while they are growing, but that this stops after they reach maturity. Let us examine the validity of this statement. The growth process of vegetation is illustrated in Fig. 6.3. The left-hand diagrams [Figs. 6.3(a) and (b)] show the case of a forest which is left untended after being planted. Of course, it represents a situation in which, after planting, the weather conditions are such that growth occurs naturally. As shown in the upper diagram [Fig. 6.3(a)], only a very small proportion of the carbon dioxide involved in photosynthesis is actually fixed, with the rate of carbon fixation reaching a peak at about 15–20 years after planting; accumulation then virtually ceases. The bottom left diagram [Fig. 6.3(b)] shows that there is no drop in the total amount of carbon accumulated, provided that no felling takes place. In this sense, the situation can be regarded as being virtually the same as subterranean storage. The difference is that in the case of afforestation it is sufficient to just dig a hole in the ground and plant a sapling. After that, the tree grows on its own and carbon is accumulated naturally by just leaving it alone. Also, it is in the form of organic materials and not carbon dioxide, although the disadvantage is that it takes quite a while to achieve. However, the time required is at most 20–30 years, which is sufficiently short in the context of the carbon dioxide problem. For this reason, the discussion so far has ignored the accumulation time.

The next problem concerns the amount of space available and the quantities that can be accumulated. Here again, Table 6.1 is useful. First, let us assume that afforestation would permit the fixation of more than 5.5 gigatons each year of the carbon which is released from fossil fuels in the form of carbon dioxide. The density of organic matter in tropical and other forests is over 200 tonnes per hectare. In tropical forests, 150 t/ha of this is above ground; afforestation in tropical grasslands would yield the same figure. On the other hand, the conversion of deserts into grassland would lead to an accumulation of several dozen tonnes per hectare. If deserts could be transformed into tropical forests, then using the previous figure of over 200 t/ha, it is apparent that it would be necessary to greenify about 25 million hectares each year. The total area of deserts is

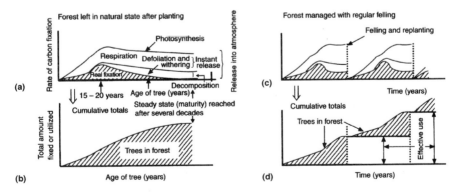

FIGURE 6.3 Forest growth and carbon fixation.

4.5 billion hectares and, if the area of grasslands is also considered, this is equivalent to the area needed to fix carbon for approximately 200 years. However, the
benchmark for the calculations is not the total amount of carbon dioxide released
from fossil fuels but the residual amount in the atmosphere (i.e. 3.5 gigatons per
year); furthermore, if the area of tropical forest that is being destroyed is subtracted from this figure (assuming that there is no further destruction of the forest), the annual rate of fixation is only 1.5–2.0 gigatons. As a result, this approach
to combating the carbon dioxide would be effective for 600 years, i.e. three
times the figure stated earlier. During this time, it is probable that sources of
alternative energy would be adequately developed, such as the use of solar energy
and fail-safe nuclear fusion.

In the final analysis, rather than thinking of forests as a means of regularly
absorbing carbon dioxide, they should be regarded as a non-steady-state storehouse of carbon which makes a large contribution to the carbon dioxide balance.
When considering the carbon dioxide problem, we tend to look only at the
amount of carbon dioxide released from fossil fuels and the amount that accumulates in the atmosphere, but we must remember that an equal (or even
greater) contribution is made by land vegetation and (as discussed in the next
chapter) the oceans.

6.5 THE COSTS OF AFFORESTATION

This section will consider several evaluations of the costs of afforestation for the
recovery and fixation of the carbon dioxide released from thermal power plants.

In California the cost of planting trees in order to remove the carbon dioxide
from urban areas was calculated at about 1 cent per kilowatt hour, whereas
calculations performed at Zurich put the cost at between 4.3 and 7.3 cents per
kilowatt hour. In contrast, calculations by Block *et al.* in the Netherlands showed
that the costs of maintaining the forest could be covered by profits resulting from
sales of lumber produced on the same land, and that only the initial costs needed
to be considered; these initial costs were estimated at approximately 0.1 cents
per kilowatt hour.

A group at Japan's Economic Planning Agency estimated that if carbon emissions were to be reduced by one tonne by decreasing Japan's domestic production of goods, this would cause an economic loss of $16,000; however, they also
calculated that if the same result were to be achieved by investments in energy
conservation in developing countries, the cost would fall to about $200. It was
estimated that the planting of tropical forests would achieve the same reduction
for a little under $100. One tonne of carbon is produced by the generation of
about 5,000 kilowatt hours of electricity, and thus this also yields an estimate of

about 2 cents per kilowatt hour. However, even if relying on afforestation, the price of crude oil per tonne of carbon equivalent is at most of the order of $200, which is equal to about half of the cost of crude oil in the early 1990s; this would translate into a price rise of several cents on a liter of gasoline.

If the cost in electricity on Japan is taken to be 5 cents per kilowatt hour, it is clear that there is a great difference from the estimated cost of using afforestation to combat the adverse effects of carbon dioxide. Of course the accuracy of the calculations is important, but there is also a considerable difference in cost depending on where afforestation is carried out. However, since the effect of afforestation is equal everywhere in the world, afforestation should as far as possible be carried out in areas which do not require the expenditure of financial resources and extra energy.

The best approaches to the problem are to prevent the destruction of tropical forests and to replant trees that have been felled; these actions are relatively inexpensive, and involve no technological difficulties. The only problems concern the system of forestry management and economics. This point (and exactly the same point applies to biomass—including biomass used for a variety of purposes such as pulp and construction) is the very one underlying the idea that the user should be the one who pays the energy tax (or environmental tax, or carbon tax); such taxes should be levied on the use of biomass in the same manner as for coal or gasoline. This energy tax should then be used to recycle the costs incurred by afforestation and forest management, and the costs arising from preventing the destruction of tropical forests in those areas in which biomass is produced. In tropical areas (which is where the deforestation is taking place) the costs of land and labor are low, and the productivity of the forests and the amount of carbon stock are both high; it is here that the destruction of forests needs to be halted, and afforestation actively pursued. It is not just that these actions have value as adequate countermeasures, but also that the advanced nations, which have themselves built their current prosperity while the forests have been destroyed, should bear an appropriate burden of the expenses.

6.6   THE ADVANTAGES OF AFFORESTATION FROM THE POINT OF VIEW OF REQUIRED NUTRIENTS

Table 6.2 shows the amounts of various elements present in land vegetation, soil and marine plankton in terms of the number of moles per mole of phosphorus. By comparing the ratio of elements in marine plankton, it is quite clear that forests contain a large amount of carbon with a small quantity of nutrient elements. Also, as mentioned above, the majority of the carbon in tropical rainforests is present in vegetation, and even in tropical regions 70% of nutrient elements such as nitrogen are contained within the soil, with most contained in the surface layers at depths

Table 6.2 Ratio of carbon and trace elements (in moles) [*adapted from* Tsutsumi, 1987].

Location	P:	N:	C:	Ca
North-east Thailand	1:	15:	2,200:	10
Broad-leaved deciduous trees in temperate zones	1:	25:	5,300:	15
Natural white fir forest, Hokkaido*	1:	20:	4,000:	10
Artificial cedar forest, Akita	1:	25:	8,700:	10
Lucidophyllus (evergreen, broad-leaved) forest, Kyushu	1:	35:	8,600:	10
Forest average	1:	25:	5,800:	10
Soil in north-east Thailand	1:	90:	1,400:	45
Soil in natural white fir forest, Hokkaido	1:	150:	2,900:	10
Soil in artificial cedar forest, Akita	1:	1,670:	48,000:	180
Coal	1:	15–300:	1,500–30,000:	1–40
Organic materials in ocean	1:	15:	120:	40
Deep ocean	1:	15:	800:	3,200
Surface ocean	0:	0:	680:	3,160

Note: Later figures in the table are calculated under the assumption that half of the organic matter is carbon.
Note: Hokkaido is the northernmost island in the Japanese archipelago;
Akita is a prefecture in north Honshu, just south of Hokkaido;
and Kyushu is the southernmost main island in Japan.

between 0 and 30 cm. In the case of temperate zones, over 90% of the nitrogen is in the soil. In addition, the nitrogen/carbon ratio in the soil is very different from that in vegetation. On the other hand, there can easily be a shortage of nitrogen in the subarctic zones (where boreal forests grow).

In general, the capacity of the soil to adsorb nutrients such as nitrogen is large, and while the seepage from fertilizers supplied to forests is very low, once the forest is destroyed, the nutrients can easily be removed by flowing water. Furthermore, when vegetation is lost from the surface, the nutrient-rich surface soil is blown away by the wind and disappears.

The reason that this topic arose is that a consideration only of the carbon dioxide problem suggests that the amount of fertilizer needed for the growth of land vegetation is not so large, and in particular that it is only a very small fraction of that needed for the growth of marine plankton. In Section 5.3, it was stated that the fixation of carbon dioxide by plankton was actually not fixation at all but should be considered as a means of technologically converting solar energy into other forms. However, this discussion clearly shows that if this plankton were to be combusted at the power plant instead of fossil fuels and thus converted into electricity, the difficulty of recovering and reusing the nitrogen and phosphorus would become a major problem. On the other hand, although the commercial production of biomass could not become a major energy source, excess materials and household garbage should be used in a positive way. At this time, there is no need to recover phosphorus and nitrogen. With the use of marine microorganisms, however, the problems are much greater.

6.7 THE IMPORTANCE OF GOOD FORESTRY MANAGEMENT

Figure 6.4 shows the degree to which the cutting down of trees is a factor in the destruction of forests. In 1986, total lumber production was $3.25 \times 10^9\,m^3$, which would amount to approximately 1 gigaton of carbon equivalent per year; also, the figure is continuing to rise today at an annual rate of about 2%. The first point that needs to be made is that in the advanced nations the production is primarily of conifers, whereas in the developing countries production is mainly of broad-leaved trees; total world production is roughly split equally between these two types.

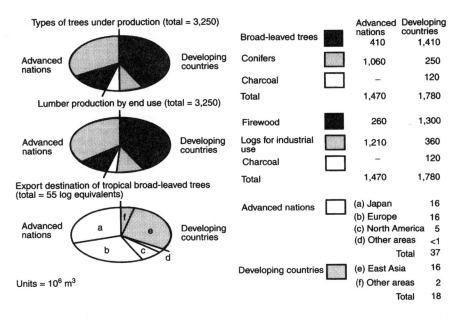

FIGURE 6.4 Worldwide forestry production and exports in 1986 [Kuroda and Nectoux, 1989].

The figure mentioned a moment ago of 1 gigaton appears to be close to the figure of 1.5–2.0 gigatons mentioned in relation to the carbon balance in Chapter 2 for the amount of carbon released as the result of the destruction of forests. However, the large majority of this 1 gigaton is basically due to the excess production of forests and is the result of a planned cycle of planting and felling; therefore it is not directly related to the release of carbon dioxide. In particular, lumber production in advanced countries is carried out in a relatively appropriate and planned manner. If carbon dioxide is actually released due to deforestation, the amount is negligible compared with tropical forests (as described in Chapter 2).

Let us now take another look at Table 6.1. Conifers, which account for almost all of the production in advanced countries, occupy an area of approximately 1.2 billion hectares—only about one quarter of the total forested land. They account for over one-third of all lumber production. Furthermore, primary production is only a small fraction of total forest. Even so, (despite the fact that the deforestation of tropical forests has become such a major problem) the felling of conifers is unimportant with respect to the carbon dioxide problem. A reconsideration of this situation therefore leads to the conclusion that lumber production should not lead to deforestation if only appropriate management methods are adopted.

Approximately 80% of the production in developing countries is for firewood or charcoal, and thus the main use is as a source of energy for cooking. The total amount of lumber consumed as an energy source (including that in the advanced nations) comes to $1.7 \times 10^9 \, m^3$, which is several hundred million tonnes of carbon equivalent. Certainly part of this is a factor in the destruction of tropical forests and the advance of desertification. On the other hand, it cannot on its own explain the destruction of the tropical forests.

Is, as has often been said recently, the export of trees to countries such as Japan one of the causes of forest destruction? If the use of wood for firewood etc. is excluded from the production totals in developing countries, it leaves only 20% of the total; of this, the amount which can be exported is merely $5.5 \times 10^6 \, m^3$, which is only about 20 million tonnes of carbon equivalent. Furthermore, the amount exported to Japan is a little short of 30% of this total, and is less than 10 million tonnes of carbon equivalent—which seems to be a rather insignificant amount.

However, although there may be no direct effect, one view holds that there is indeed an indirect effect. The argument is that in order to export the wood, it is necessary to fell top-grade trees, but that such trees account for only a small percentage of the total number of trees, and that in order to cut down these high-quality trees it is necessary to also fell many other trees. Furthermore, the roads that are needed in order to carry out this felling lead to the development of these regions; this includes the influx of people that accompanies construction of the roads, the starting of fires, and the adoption of slash-and-burn agriculture. After roads are opened to traffic, development then proceeds in the area around them.

In other words, the number one cause of the destruction of tropical forests is migrant farmers who make the land arable by using slash-and-burn techniques.

One estimate even suggests that this accounts for over half of all the deforestation. This is not to say that slash-and-burn agriculture should be banned. From a long time ago people have made their living in this way, but the situation remained constant as long as there was no change from the regular pattern (i.e. no large-scale unplanned expansion), and provided there was the repeated application of the cycle of agricultural land, abandonment, return of plant life, reforestation, and slash-and-burn agriculture. The problem rests with the erratic implementation of the technique, and is deeply interwoven with the problems of the population explosion and the availability of food supplies.

Nutrients are certainly lost from the ecosystem during lumber production activities, but for a time the amount of nutrients in the soil increases due to the decomposition of the fallen leaves and branches. Furthermore, although the slash-and-burn techniques mean that some of the carbon, nitrogen and phosphorus is lost to the atmosphere, the minerals contained in the organic matter undergo decomposition and conversion to an inorganic form, and since this is a form which is easy for plant life to take up, the land outwardly appears to be very fertile. This is both the strong and weak point of slash-and-burn agriculture. As time passes, the ecosystem gradually becomes replenished with the elements which were lost, such as phosphorus. There is no problem provided that the plant life remains while this process is taking place, farming activities are halted, and plenty of time is allowed for the forests to return. However, once the fertility of the land is squandered, recovery is difficult, and the organic matter is converted into inorganic materials, the water content of the soil disappears, and the remaining nutrients are lost as a result of erosion. The forest disappears, and a further problem arises in the form of soil salinity in the cultivated land since evaporation in the dry season takes place from the surface of the land, resulting in the salt content of the water accumulating in the surface of the earth. Under these conditions it is difficult for forests to recover, and the end result is that the land becomes abandoned. As shown in Table 6.3, a large proportion of the tropical forest that has disappeared is today not used for agricultural purposes but just left to stand as it is.

Table 6.3 Usage of the lost tropical forest [International Intergovernmental Panel on Climate Change (IPCC), 1990].

Type of land	Latin America	Africa	Asia
Agricultural land*	45%	62%	90%
Grazing land	17%	−19%	−1%
Abandoned land	38%	57%	11%

*This figure includes land used for transient slash-and-burn agriculture.

Let us briefly recap the situation. If only the tropical forests were managed in an appropriate fashion, the present level of lumbering activities could probably be maintained without the destruction of the forests themselves. It is thus hoped that fresh employment opportunities will develop so that the need for slash-and-burn agriculture will be obviated.

6.8   THE USE OF FIREWOOD AND SUBSTITUTION WITH ALTERNATIVE ENERGY

As mentioned above, the entire question of deforestation is extremely complex. There are several causes of the problem; in addition to the activities of the timber industry and the felling of trees in slash-and-burn agriculture, deforestation also results from the cutting down of trees for use as firewood. Let us now consider the use of firewood as a source of energy.

The amount of firewood used is growing year by year and, as mentioned earlier, accounts for 80% of lumber production in developing countries. There would be no problem if the only wood used was the excess production during the process of regenerating mature forests; indeed this is how biomass used to be used. However, the problem exists because the use of firewood has become a cause of deforestation and desertification.

There is a common perception that the substitution of firewood with fossil fuels such as coal would be linked to an increase in carbon dioxide emissions. However, if the recent pressures arising from the population explosion mean that even young trees (which should go on to develop into forests and thus be capable of fixing a relatively large amount of carbon) are used as firewood, and if in addition it is accepted that this forms a major cause of deforestation and desertification, then the use of coal would instead actually be helpful in combatting the carbon dioxide problem. The area of land which is being deforested and turned into desert due to the collection of firewood is said to amount to 2 to 3 million hectares each year. Even if the amount of carbon released as a result of this is taken to be half of the difference between the figures in Table 6.1 for the amount of carbon contained in woodland and shrublands and the quantity in the desert ($= 70$ tonnes per hectare, which is equal to the difference between woodland/shrublands and the tropical grasslands, or the difference between tropical grasslands and desert), this would mean a total release of carbon of nearly 0.2 gigatons, which is over 0.2 gigatons of coal equivalent. That is equivalent to nearly half of the total firewood use given in Table 3.4. It might be thought that only a relatively small proportion of firewood use invites the destruction of the ecosystem; even so, it would be possible to cut the total use of firewood by half if it were acceptable for the amount of carbon dioxide released because of the destruction of vegetation that is entailed in the collection of firewood to be emitted as a result of the use of coal and oil, and if this amount of firewood were then to be replaced by the use of coal and oil.

For various countries, Fig. 6.5 shows the forms of energy used and the proportion of total energy use accounted for by cooking, and also the per capita consumption of energy for cooking. In those areas of developing countries where firewood and charcoal are used, the amount of energy consumed for cooking is relatively high, and the efficiency is low. In the case of firewood, this is because the firewood usually contains some water, and energy is needed to make this evaporate; in addition, some fuel is already consumed before actual cooking takes place, and some continues to be burned after cooking is finished. When these points are also taken into consideration, the problem of firewood use looms even greater. In certain cases, even if it is regarded as natural that coal is useful in preventing desertification, it is possible that the use of coal can play a considerable role in the carbon dioxide problem; oil and gas are even more efficient. However, when considering acid rain etc., it is necessary to adopt measures such as requiring the pre-treatment of coal and oil in order to reduce the sulfur content.

Furthermore, in countries which have a problem with deforestation and desertification resulting from the chopping down of trees for firewood, it is easier to make use of natural energy such as solar energy than it is in countries like Japan. Even if the energy payback time for photovoltaic cells in Japan is taken to be 10 years, this could be reduced by half if the cells were located on the edge of the deserts. Energy efficiency would rise just by moving the photovoltaic cells from Japan to the desert. Even if one did not go to the extent of using photovoltaic cells, an extremely important contribution to the carbon dioxide problem

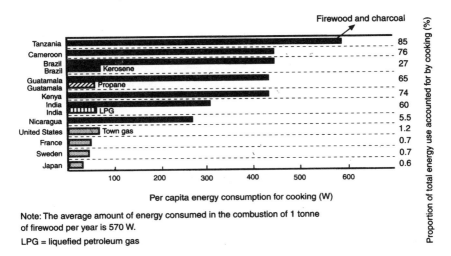

Note: The average amount of energy consumed in the combustion of 1 tonne of firewood per year is 570 W.

LPG = liquefied petroleum gas

FIGURE 6.5 The per capita energy consumption for cooking in various developing countries [Kammen, 1993].

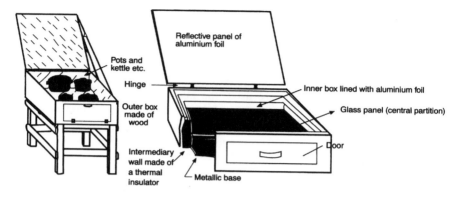

FIGURE 6.6   The structure and operation of a solar oven [Kammen, 1993].

could be made by using solar ovens to supply hot water, and by using convex lenses and concave mirrors to concentrate the sun's rays so as to provide a source of heat for cooking. With just a sheet of aluminum foil affixed to the surface of a box instead of mirrors (Fig. 6.6), it only takes 50 minutes to boil two liters of water.

Even if coal was for a time used in Japan, substitution with solar energy in areas where this can be done relatively simply would play a useful role in a global problem such as that surrounding carbon dioxide. It may be extremely difficult to evaluate this in monetary terms, but the investment of just a small amount of money can produce a considerable reduction in carbon dioxide. Although at present it is possible for volunteers to send solar ovens on an individual basis, the provision of such aid should definitely not be just left up to individuals who are concerned about the situation; it is to be hoped that some new concept can be devised which will be consistent with business logic, for example the introduction of a new system which involves the implementation and use of a carbon tax or energy tax. This would indeed help strategies aimed at promoting the introduction of photovoltaic cells and increased afforestation and greenification.

6.9   THE CONSTRUCTION OF WOODEN HOUSING AS A MEANS OF CARBON FIXATION

Forestry resources can be regenerated if the cycle of felling and afforestation is repeated in a planned manner; also, it may be possible to use up to several percent of the energy supplied by photosynthesis in the form of lumber and biomass. Let us now consider the value of wood for housing and furniture etc. and the possibility of carbon fixation.

First let us attempt to grasp the situation intuitively. Let us take the case in which the forest is destroyed, the land is cleared, and a house is then built in the same place as the forest once stood. The parts which could not be used as building materials are returned to the soil; these and the organic matter contained within the soil then undergo decomposition, with the production of carbon dioxide. Let us now consider an alternative case whereby the forest is cleared under constant supervision in such a way that the forest is not destroyed, and that this part is reforested. What would the situation be now if that wood were to be used as a building material? As long as the building was in use, the carbon that was being used would have been removed from the atmosphere and undergone fixation.

On the other hand, what would happen if other building materials were used? Of course, energy would be expended in the production of these other materials. Even in the case of wood, energy is needed for cutting down the trees, for processing the wood, and for transportation. Figure 6.7 shows the results of calculations of the energy needed for the materials used in various building methods and for transportation etc., assigning a negative value for the carbon which is fixed in the wood that is used. In comparison with steel, reinforced concrete and steel-reinforced concrete, not only is the amount of energy used in wood construction smaller, but also the wood acts as a means of carbon fixation. Thus, if forestry resources could be maintained, the construction of wooden houses could make a contribution to solving the carbon dioxide problem.

Does that mean that by using the forests, cutting down the trees and preserving the wood (preserving it either as wooden housing, or else in arctic and desert

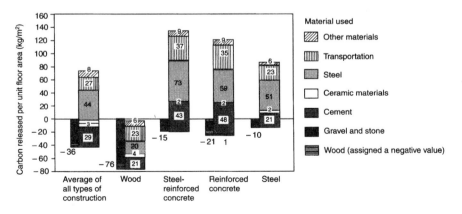

FIGURE 6.7 Carbon dioxide released from building materials and fixation by wood [Sakai, 1993].

areas), it is possible to fix all the carbon dioxide that is released from fossil fuels? It was mentioned earlier that only about 20% of the net primary production of organic matter is actually used for tree growth. In forests which are artificially managed productivity is slightly higher, with the annual average rate of wood production being 1–2 tonnes of carbon per hectare. If the carbon dioxide emissions from fossil fuels (5.5 gigatons of carbon) are divided by this figure, the necessary area of forest is calculated to be several billion hectares. This figure is comparable with the total area of forested land (including the Amazon etc.). Furthermore, if it is postulated that this would be used to build wooden houses on the land, it would necessitate a vast amount of land (unless wooden high-rise buildings were constructed). Also, it would be nonsensical to deliberately pile up all the wood. In actual fact, the majority of the wood is used in ways that make it impossible to preserve, such as for firewood and paper, thus rendering the entire idea unrealistic.

6.10   THE PROBLEM OF DESERTIFICATION, AND THE ENERGY NEEDED FOR GREENIFICATION

As mentioned earlier, carbon dioxide is released due to deforestation, but if afforestation is carried out, it is fixed as carbon. However, once a mature forest is formed it can no longer act as a sink for carbon dioxide absorption. Therefore the question of the extent to which afforestation can make a contribution to the problem depends on estimates of the current rate of deforestation, the area available for afforestation, and the amount of carbon fixation per unit area. Provided the evaluation is performed over the long term, the rate at which the forest sequesters carbon dioxide is not a major problem.

If the vegetation on the land surface were changed, i.e. if a desert which contains no greenery were converted into forest, 200–260 tonnes of carbon could be fixed per hectare (Table 6.1). Alternatively, if the deserts were transformed into grasslands, or if the grasslands were converted into forests, these processes would each contribute about half of this amount. The area covered by desert is believed to be large enough to play an extremely important role in the carbon dioxide problem. That being the case, why is greenification not initiated immediately? In actual fact, the questions of whether deforestation can be halted and whether afforestation is possible depend not only on the special characteristics of the areas and regional differences such as geological features and climate, but also depend greatly on factors such as sociology, economics, the ethnic group to which the people there belong, cultural characteristics, and the educational level of the people. Nevertheless, in the true desert areas which have not previously been used by human beings, such problems are not serious and are merely a question of technology. Of course there are problems with agriculture, but it should be possible to technically solve the difficulties if only water could be supplied.

In places where there is sufficient rainfall or where water can be obtained from rivers etc., minimal energy is required for collecting the water, transporting it, and using it for greenification; this should also pose no great technological difficulties. Vast areas could be greenified using this strategy. However, a problem soon becomes evident when a distribution pattern of global precipitation (e.g. the one by Japan's National Astronomical Observatory, 1990) is compared with a map of global vegetation (Fig. 6.8); namely, the deserts have become deserts because of the lack of rainfall.

Accordingly, let us consider a case study of an area in which the sparsity of rainfall has prevented the growth of a forest, but where fresh water is obtained artificially from salt water. (Of course this discussion would not be applicable to desert areas which possessed no salt water or in highland regions where the transport of sea water would be difficult.)

It is difficult for a forest to develop in areas such as deserts where the annual rainfall is less than 100 mm, and it is thus necessary to think of some method of irrigation. Let us now consider places where precipitation is close to zero and no rivers are in the vicinity, but under the assumption that sea water or salt water is readily available, and that the fresh water needed for greenification can be obtained only by desalination.

First, let us consider the production density of net primary production in Table 6.1. In the case of forests, it amounts to approximately 5 tonnes of carbon equivalent per hectare per year. If the dry matter is taken as about double this, this is equivalent to $1 \, kg/m^2/yr$. The amount of water believed to be necessary is 600 times the quantity of dry matter, i.e. $600 \, kg/m^2/yr$. This would be equivalent to an annual rainfall of approximately 600 mm. Figure 6.9 represents the desalination of sea water by reverse osmosis and the energy balance involved when the water is used for greening the desert. The quantity of electricity needed to prepare just this amount of water is equal to $33 \, MJ/m^2/yr$. On the other hand, even if all the biomass from the forest were temporarily combusted, the thermal energy obtained would be at most half that amount; also it would be in the form of heat, not electricity.

Another method of preparing fresh water is by evaporation, also known technically as multi-step flash evaporation, but this requires a massive $150 \, MJ/m^2/yr$ of heat. Numerous desalination plants employing this method or reverse osmosis exist in the Middle East, but the use there of the water for greenification actually involves some serious drawbacks from the point of view of the carbon dioxide problem.

Figure 6.10 shows an example of desalination equipment which uses solar thermal power and which is actually used in arid regions. In a type of greenhouse, the water vapor which is driven off from the saline water cools upon coming into

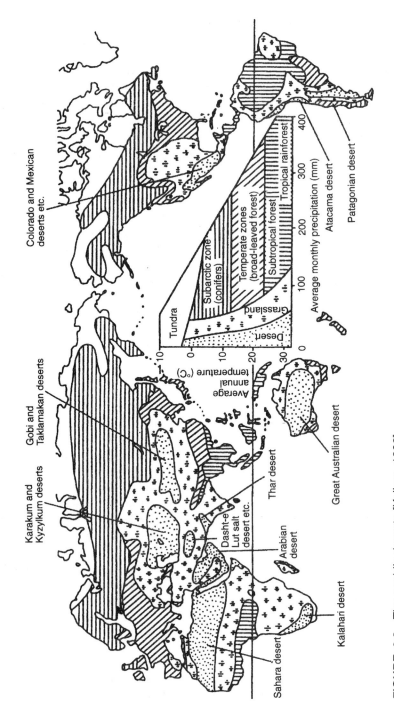

FIGURE 6.8 The world's deserts [Kojima, 1992].

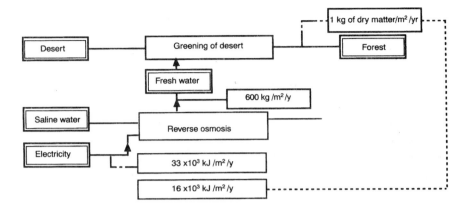

FIGURE 6.9 Energy balance when sea water is desalinated by reverse osmosis and used for greening the desert [Matsumura and Kojima, 1991].

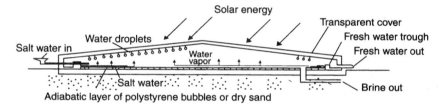

FIGURE 6.10 Basin-type solar still [Matsumura and Kojima, 1991].

contact with the roof, and condenses. The condensed water is then collected using open conduits. The maximum thermal efficiency is about 30–40%, which means that the simple equipment has a relatively high efficiency. Figure 6.11 shows the situation in which this equipment is set up in the desert. Even if the water has to be raised to 40 meters above sea level, this does not require so much energy, and since solar energy is being used, there is no problem with the energy balance. In order to create one hectare of forest, the area covered by the desalination equipment would need to be a minimum of half this. It would thus be impossible to convert more than one-third of the desert into forest.

It is evident from Fig. 6.12 that it is extremely difficult to greenify the desert, but at the same time, if it is recognized that the carbon dioxide problem is both enormous in scale and also a matter of major significance for the human race, then contrary to expectations it becomes a realistic proposition. The water shortage which occurs in the dry season could be offset by using the equipment in conjunction with photovoltaic cells, or else by increasing the efficiency by making it a multi-step procedure that employs methods of heat recovery.

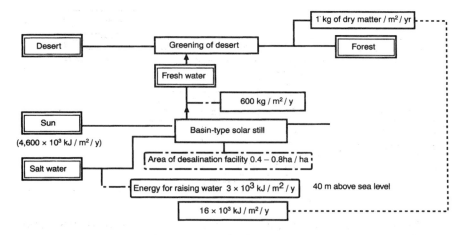

FIGURE 6.11 Energy balance and scale of basin-type solar still and subsequent use of the water for greening the desert [Matsumura and Kojima, 1991].

This section looked at the production of water in the desert, but there should be locations where the conditions are not so harsh, and where greenification and afforestation could be accomplished more easily. Initial projects should concentrate on areas where the desired outcome can be achieved relatively easily, and these could then gradually be applied on a wider scale.

6.11 METHODS OF EVALUATING AFFORESTATION

It should be noted that none of the techniques proposed here require the production of energy, and for this reason energy efficiency itself is not a problem. Energy which cannot be used by humans is used by trees, which in turn convert carbon dioxide into carbon.

In order to absorb a given amount of carbon dioxide, this requires less fertilizer than is needed for ocean fertilization (see Chapter 7), and as previously mentioned the fertilization problem involved in afforestation of the desert is not serious in comparison with the water problem. At the beginning, afforestation should be performed in those areas in which results can most easily be achieved (i.e. where the use of a few special techniques for replanting can produce a significant difference), and not in locations where there is no water.

The question of environmental destruction is the most vital issue, not just from the point of view of afforestation, but also from that of the carbon dioxide problem. The question of whether to implement the above system depends on whether we adopt policies to meet identified needs. Consequently, it is meaningless at the present time to debate the economic aspects.

However, it is necessary to adequately examine techniques from the viewpoint of energy to see whether they are truly appropriate. Let us consider some functional evaluation for afforestation (and it should also be possible to establish similar procedures for other countermeasures). The first question concerns the time scale that should be considered for the technology. In one hundred years, or a few hundred years, alternative forms of energy should be available; for this reason, a period of one hundred years appears to be an appropriate length of time.

Despite the earlier comments about the need for evaluating the total accumulation over time, this can be ignored if virtually no increase is observed even after the passage of one thousand years. However, if this should occur at a slightly faster rate, there is no problem with evaluations on the basis of cumulative totals. Consequently an evaluation of the average carbon stock over a 100-year period would also take into consideration the possibility of slow growth.

The first thing that must be considered is the ratio of the effective amount of carbon fixation to the energy that needs to be input. The greater this value, the greater should be the speed with which the approach is implemented. The effective amount of energy can be calculated by subtracting the energy input from the energy equivalent of the fixed carbon (the carbon-to-energy conversion can be performed on the basis of the heating value/carbon ratio of a fossil fuel, e.g. coal). If this value is positive (i.e. there is a gain in energy or in net carbon fixation), the technique would be beneficial, and the greater the value, the greater the certainty that the technique would be effective.

The amount of carbon fixation is the average amount of carbon retained in the ecosystem (including the soil) over a period of one hundred years after the policy has been implemented minus that which was present before the policy was begun. When there is, for example, a repeated regular cycle of felling and replanting in a man-made forest, the average value over the 100-year period is taken to be roughly the average between the two states (i.e. of mature forest being present, and felling having just been completed). Then the value is converted into an energy equivalent as described earlier.

The net energy input is the cumulative total of energy which must be input over the hundred years for fulfillment of the policy (represented in terms of fossil fuel equivalent) minus the cumulative total of energy obtained over the same period of time from the forest (as firewood etc.). Besides the energy needed to produce water, it is also necessary for the first of these terms to include the energy needed to construct the facilities for both manufacturing and storing the water. Also, the greater the effective amount of carbon fixation per unit area (i.e. the total of fixed carbon minus the energy input, expressed in terms of carbon equivalent), the more effective will be the replanting program.

The second point that should be addressed is the question of what percentage of land vegetation would be included in this policy over the one hundred years if this technique were used as the sole means of resolving the carbon dioxide problem. That is to say, it is essential to accurately evaluate the possibility of continuing the policy. In the case of afforestation, the end point would come when all the earth's vegetation were replaced by forest. If either the total carbon equivalent accumulated in the atmosphere worldwide over the hundred years, or else the amount released from fossil fuels, were divided by the effective carbon fixation per unit area, and then this value further divided by the total global area covered by vegetation to cope with the situation, this would permit calculation of the percentage of land vegetation referred to at the start of this paragraph. If the figure were to exceed 100%, this would mean that the policy could not be implemented successfully within a one-hundred-year time scale. In the case of desert greenification, as mentioned earlier, only a small fraction could be greenified in just one century.

The final point that must be evaluated is the amount of manpower needed to carry out the afforestation. If the number of workers per unit area needed to carry out the policy over one hundred years were multiplied by the annual worldwide per capita amounts of carbon equivalent either accumulated in the atmosphere or released from fossil fuels, and this then divided by the effective amount of carbon fixation per unit area, this would yield the percentage of the world's population that would need to work for the implementation of the policy. A value of 0.1 would mean that one out of every ten people would be involved in afforestation. If the number of people needing to work in this field were not so large, there would be no problem. However, if a large number of people quit their jobs in agriculture, commerce and industry in order to devote themselves to afforestation, there would be a corresponding decrease in the amount of production carried out by the human race. There is also the question of the extent to which people would value the afforestation work. However, if all people had to work to implement afforestation, human society would collapse.

These evaluations are only preliminary ones and can by no means be regarded as fully adequate. Certainly, there are various complex and economically important benefits associated with afforestation such as the use of amenities, flood prevention, the prevention of soil erosion, water resources, and climatic changes. In the distant past, biomass produced from forests supplied virtually all the energy consumed by human beings. Indeed it could be said that forests were the final fortress for the human race. However, when thinking of these forests in terms of countermeasures against carbon dioxide, then as was apparent in the case of greenification of the deserts in the Middle East, not everything is favourable. Hard-headed evaluations such as the one above are absolutely vital.

In conclusion, it is clear that afforestation and greenification must be gradually promoted in the interests of all human beings, beginning with those areas where they would prove to be the most useful.

References

Food and Agricultural Organization of the United Nations. 1992: The Forest Resources of the Tropical Zone by Main Ecological Regions by Forest Resources Assessment, 1990 Project. Cited in Ohsumi, Y. 1993. In Kojima, T. (ed.). Seminar on Global Environment, Vol. 5, p. 58, Ohm Pub., Tokyo, Japan (in Japanese)

International Intergovermental Panel on Climate Change (IPCC) 1990. *Report of the Subgroup on Agriculture, Forestry and Other Human Activities* (IPCC Working Group III). Cited in Ohsumi, Y. 1993. In Kojima, T. (ed.). Seminar on Global Environment, Vol. 5, p. 58, Ohm Pub., Tokyo, Japan (in Japanese)

Kammen, D. 1993. Nikkei Science (Special Issue for the Second Earth Awards) (5-S), 6 (in Japanese, original in English)

Kojima, K. 1989. 3, 14 Energy Review (in Japanese)

Kojima, T. 1992. In Komiyama H. (ed.). *Introduction to Chemical Technology for the Global Environment*. Ohm Pub., Tokyo, Japan (in Japanese)

Kuroda, Y. and F. Nectoux. 1989. *Timber from the South East*. World Wide Fund for Nature. International, Switzerland. (Citation from the Japanese edition, 1989, Tsukiji Pub., Tokyo, Japan)

Matsumura, K. and T. Kojima. 1991. *Journal of Arid Land Studies* 1(1), 73 (in Japanese)

Oikawa, T. 1989. *Modern Chemistry* 11, 61 (in Japanese). Cited in Kojima, T. 1990. Mol 28(5) 56 (in Japanese)

Sakai, K. 1993. Proceedings of The 8th National Congress for Environmental Studies, p. 2. The Mining and Materials Processing Institute of Japan, Jan. 20th 1993, Tokyo, Japan. (in Japanese)

Tsutsumi, T. 1987. *Material Cycle in Forest.* p. 27, 30. Tokyo Univ. Pub., Tokyo, Japan (in Japanese)

Tsutsumi, T. 1989. *Forest Ecosystems.* p. 166. Asakura Pub., Tokyo, Univ. Pub., Tokyo, Japan. (in Japanese)

Whittaker, R. H. and G. E. Likens. 1973. In Woodwell, G. M. and E. V. Pecan (eds.). *Carbon and Biosphere.* US Atomic Energy Commission, p. 281

Woodwell, G. M. 1987. *Science* 11, 20 (in Japanese)

Woodwell, G. M., W. A. Whittaker, G. E. Likens, C. C. Delwiche and D. B. Botkin. 1978. *Science* 199, 141

Yoda, K. 1982. *Geochemistry* 16, 78 (in Japanese)

7 THE ROLE OF THE OCEANS

7.1 INTRODUCTION

The oceans have a great capacity for absorbing carbon dioxide (Fig. 2.8) and act as a major sink even though they are unable to absorb all the carbon dioxide released from fossil fuels. Computer simulations of the absorption of carbon dioxide have been imprecise and unreliable, which has been attributed to the oceans possibly acting as a missing sink. If the oceans did absorb all the carbon dioxide which is unaccounted for, they would be removing 3.5–4.0 gigatons of carbon each year. A mere doubling of this amount would resolve the carbon dioxide problem.

Under steady state conditions, the absorption of carbon dioxide by the oceans would equal the amount released from the oceans into the atmosphere, i.e. there would be a balanced exchange. However, the oceans in fact display a net absorption of carbon dioxide, demonstrating that the increase in carbon dioxide concentrations has disturbed the steady state. The carbon dioxide which has been released from fossil fuels as a result of human activity (particularly since the industrial revolution) has been dispersed both into the atmosphere and into the oceans, with the amount of carbon dioxide dissolved in the oceans actually being about 60 times that contained in the atmosphere (Fig. 2.7).

At equilibrium, the further absorption of carbon dioxide changes the nature of sea water, which means that the capacity of sea water to absorb carbon dioxide is theoretically limited to only about 6 times that of the atmosphere; even so, the actual figure for the oceans should be considerably greater than 2.0–2.5 gigatons per year (the value indicated in Fig. 2.8). On the ocean floor more than ten thousand times as much carbon dioxide exists in the form of calcium carbonate than is present in the atmosphere. When one also considers that the limestone on the sea bed reacts with the carbon dioxide to produce bicarbonate ions in the sea water, it would seem that in theory the oceans have a virtually limitless capacity to absorb carbon dioxide.

However, this is not the case in practice. Although the surface layers are well-mixed, mixing is very poor at depths between 100 and 1,500 m. Consequently, mixing between the deep ocean and the surface layers occurs at an extremely slow pace; the dissolution of the calcium carbonate at the ocean floor takes place even more slowly. This greatly limits the uptake of atmospheric carbon dioxide by the oceans.

A marked discrepancy appears to exist between the increase in anthropogenic carbon dioxide and the sum of the residual amount in the atmosphere plus the

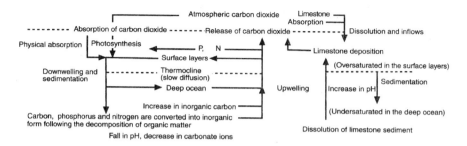

FIGURE 7.1   Circulation of carbon and sea water in the oceans.

amounts that are removed by known processes. This therefore suggests the existence of some 'missing sinks', and it is thus thought that mixing may occur as a result of mechanisms which have yet to be clarified.

All the deep waters of the oceans undergo a global circulation; the water sinks in the north Atlantic Ocean and reaches the Pacific Ocean approximately one thousand years later. The sea water then returns to the surface layers, and follows approximately the same route in reverse. As the water circulates in this way, the carbon (and hence carbon dioxide) is involved in various exchange and transformation processes. The overall carbon cycle and the transfer of materials due to the global circulation of sea water and exchanges between the surface layers and the deep ocean are depicted in Fig. 7.1.

The first half of this chapter will discuss the above issues in greater detail. The latter half of the chapter will then provide examples of how the oceans could be utilized to absorb carbon dioxide from the atmosphere, and will also evaluate these various possibilities.

7.2   THE BEHAVIOR OF SEA WATER

Since policies that affect the seas must be based on a thorough understanding of the behavior of sea water, let us first summarize current knowledge concerning the oceans.

The total surface area of the world's oceans is approximately 360 million square kilometers (about 71% of the earth's surface); around half of this is accounted for by the Pacific Ocean, followed by the Atlantic and Indian Oceans. Sea water accounts for 94% of all surface water in the world ($1.36 \times 10^9 \, \text{km}^3$); the average depth for all oceans is 3,800 m, with the average depths of the Atlantic and Pacific Oceans respectively being 3,300 and 4,000 m. The difference between the depths affects choices concerning measures to counteract the effects of carbon dioxide.

Sea water undergoes a global circulation, with the water sinking to the abyssal depths and then later reappearing at the surface. The main area in which sea water sinks is the ocean off the coast of Greenland in the North Atlantic Ocean, with the amount involved believed to be several hundred trillion tonnes per year. A large quantity of sea water also sinks in the Antarctic (believed to amount to nearly one thousand trillion tonnes); however, the large majority of this is water that has travelled via the deep ocean from Greenland and rises to the surface before again falling back to the depths.

The reason for cold water sinking into the deep ocean at high latitudes is that the density of sea water rises as the temperature drops and continues to do so even at a temperature below 4°C (4°C is the temperature at which fresh water displays its maximum density); furthermore, sea water does not form ice even when the temperature falls below 0°C. Due to energy gained from the rotation of the earth, the water which sinks flows to the west as deep water, follows the western borders of the ocean, and finally heads toward the equator. When it arrives there, some of the water starts moving east and at the same time rising slowly. The rate at which it rises is of the order of several meters per year, although it ascends slightly more quickly when it strikes the continents. What might be called the final exit from the deep ocean is in an equatorial region about 200 km wide; here, the water flows in an easterly direction and rises at the rapid rate of several hundred meters per year. It then flows northwards in the surface layers (Fig. 7.2). In the areas where the flows rise to the surface, it is the deep water that is upwelling; this produces various special characteristics, and it is these that hold the key to the countermeasures for combating the problems caused by carbon dioxide.

Consideration must also be paid to factors such as evaporation. Annual evaporation from the oceans amounts to $450,000 \, \text{km}^3$ (450×10^{12} tonnes) of water vapor; division by the total volume of the oceans reveals that it would take about

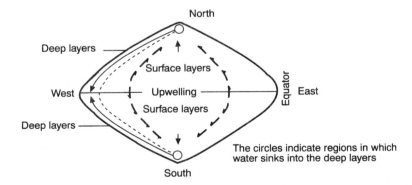

FIGURE 7.2 Principles governing the circulation of ocean water [Tomodo, 1983].

3,000 years for all the water to evaporate even if none were returned to the ocean. However, evaporation and precipitation do of course occur all the time, but the sites are unevenly distributed; as a result, the distribution of rivers is also unequal, forming another means by which water is exchanged between the oceans. Generally speaking, each year about 10^4km^3, or 10×10^{12} tonnes, is transferred in this way from the Pacific and Arctic Oceans into the Atlantic and Indian Oceans.

As stated, Arctic water sinks in the Atlantic Ocean off Greenland, from where it flows to the Antarctic; it then goes through the rising and sinking process described earlier before arriving in the Indian and Pacific Oceans (Fig. 7.3). As a result of these flows between the Pacific and Atlantic Oceans, there is a big difference in the age of the sea water (here, age is defined as the time since the water was at the surface) (Fig. 7.4). The age of the deep water of the Atlantic Ocean is several hundred to one thousand years, whereas the deep water in the Pacific is between one and two thousand years old.

Now let us consider how the mixing and diffusion of sea water varies with depth, and how vertical transfer takes place. Heat from the sun raises the temperature of the surface layers (especially at low latitudes). The surface layers are well mixed as a result of agitation by wind, and hence are referred to as either the "surface mixed layer" or the "wind-mixed" layer. The depth of the surface mixed layer varies between less than one hundred meters to several hundred meters (with an average depth of 100–150 m).

In the absence of eddy diffusion, the waters below the surface water are not warmed by convective processes. Consequently a temperature discontinuity exists directly below the surface layers. This region, which has a steeper vertical

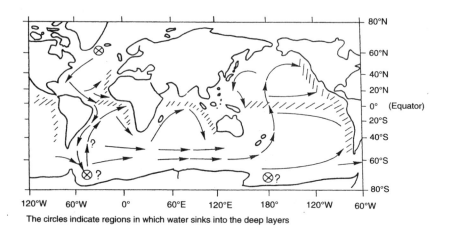

The circles indicate regions in which water sinks into the deep layers

FIGURE 7.3 Global circulation of sea water [Kojima, 1990, *adapted from* Stommel and Aron, 1960].

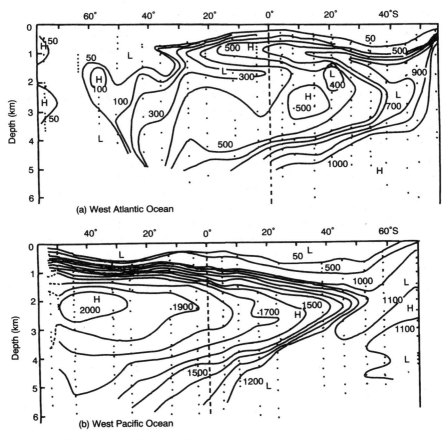

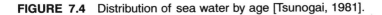

All figures within the diagrams indicate age in years

FIGURE 7.4 Distribution of sea water by age [Tsunogai, 1981].

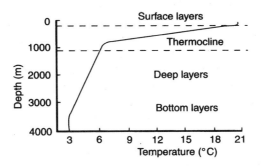

FIGURE 7.5 Temperature distribution of sea water [Turekian, 1968].

temperature gradient than the ocean layers both above and below, is known as the "thermocline". Its depth averages from 100 m to greater than 1,000 m, with the maximum occurring in mid-latitudes. Mixing is poor in this region; water and carbon dioxide is transferred only very slowly across the thermocline, which thus acts as a barrier to downward mixing. A certain degree of mixing does occur in deep sea water, although not to the same extent as in the surface layers.

The situation in the ocean could thus be simply summarized as a well-mixed surface layer above a region (the thermocline) of very poor mixing; below this is another comparatively well-mixed deep layer. This view underlies the construction of the "two box" model (introduced in Fig. 2.16), in which the surface ocean and the deep ocean are represented as well-mixed (but separate) boxes.

Although the previous discussion considered the general global situation, it needs to be borne in mind that the circulation of deep ocean water is also strongly affected by the topography of both the land surfaces and ocean floors and regional differences in the salt concentrations due to factors such as the amounts of precipitation, the inflows from rivers, and the amounts of evaporation. This means that the flow of deep ocean water can also be influenced by changes in the patterns of precipitation or by human activities such as the construction of large-scale canal projects.

7.3 DISSOLUTION OF CARBON DIOXIDE IN SEA WATER

$$CO_2 + H_2O = H_2CO_3 \tag{1}$$

$$H_2CO_3 = H^+ + HCO_3^- \tag{2}$$

$$HCO_3^- = H^+ + CO_3^{2-} \tag{3}$$

Equations (1)–(3) describe the dissolution of carbon dioxide in sea water. The majority of this carbon dioxide exists in sea water in the form of bicarbonate ions (HCO_3^-) rather than as carbonate ions (CO_3^{2-}). When calculating the specific concentrations of the various ions, it must be remembered that water has a dissociation equilibrium, and that there must be overall electrical neutrality (i.e. the number of positive and negative charges must be equal). The absorption of carbon dioxide would result in higher concentrations of carbonic acid, which might seem to infer acidification of the oceans. However, the sea water at the surface remains alkaline because carbonic acid is only a weak acid, and the concentration of positive strongly alkaline ions is in excess relative to the concentration of negative strongly acidic ions [The excess concentration is referred to as alkalinity and is roughly equivalent to ($Conc._{HCO_3^-} + \{2 \times Conc._{CO_3^{2-}}\}$), as shown in Table 7.1]. Other points to consider are that these equilibrium constants are

Table 7.1 Alkalinity and the concentrations of various ions in sea water [Broecker, 1979].

Form of carbon dioxide / Sea water	Undissociated CO_2 (mol/m³) [1]	Bicarbonate ions, HCO_3^- (mol/m³) [2]	Carbonate ions, CO_3^{2-} (mol/m³) [3]	Total carbon dioxide (mol/m³) (=[1]+[2]+[3])	Alkalinity* (mol/m³) (=[2]+2×[3])
Warm surface water	0.01	1.65	0.35	2.01	2.35
Cold surface water	0.01	1.95	0.20	2.16	2.35
Deep layers of the Atlantic Ocean	0.015	2.10	0.15	2.26	2.40
Deep layers of the Pacific Ocean	0.02	2.35	0.10	2.47	2.55

*Here, alkalinity is the excess concentration of positive strongly alkaline ions relative to the concentration of negative strongly acidic ions, and is roughly equivalent to (Conc.$_{HCO_3^-}$ + {2 × Conc.$_{CO_3^{2-}}$}).

influenced by factors such as temperature, and also that there is a slight difference between pure water and sea water.

If the reactions are assumed to be in a state of equilibrium, and calculations are performed using current atmospheric concentrations and the above alkalinity values, various things become apparent. First, the calculated results are approximately the same as those actually observed in the surface waters of the oceans. The concentration of total dissolved carbon dioxide (which includes undissociated carbonic acid, plus bicarbonate and carbonate ions) is 2.0–2.2 mol/m^3; in surface water, this concentration decreases with a rise in temperature. The relative composition of the various forms of carbon dioxide also vary with temperature; in warm surface layers, 80% is in the form of bicarbonate ions, a rate which rises to 90% in cold surface layers (the amounts of the various ions in sea water are listed in Table 7.1). The world's oceans are slightly alkaline; the pH* of the sea water is 8.1–8.3.

The above facts serve as the basis for evaluating and predicting the feasibility of ocean disposal as a means of counteracting the adverse effects of carbon dioxide, and for assessing the changes in both the oceans and the atmosphere produced by an increase in atmospheric carbon dioxide.

Let us now consider the situation in which the atmospheric concentration of carbon dioxide increases from 1 to $(1 + \partial)$, where ∂ is much smaller than 1. The amount of dissolved carbon dioxide (in the form of H_2CO_3) also increases by $(1 + \partial)$; some of this then dissociates into H^+ and HCO_3^- ions [Eq. (2)]. However, this infinitesimally small change has a considerable effect on the pH of the sea water, which then suppresses further dissociation. As a result, there is almost no change in the amount of HCO_3^- ions present in sea water. Consequently, there is effectively no increase in the amount of total dissolved carbon dioxide $(=H_2CO_3 + HCO_3^- + CO_3^{2-})$ [to be precise, it actually rises by $(1 + \partial/\beta)$, where β, the buffer factor, is approximately equal to ten].

Next, let us consider the relationship between carbon dioxide in the surface water and in the atmosphere. Overall, these have arrived at an approximate equilibrium. However, as was mentioned earlier, there is an upwelling of water at the equator, as a result of which carbon dioxide is released into the atmosphere (this will be discussed later). On the other hand, as the sea water moves northwards, the water temperature drops, and as a result the solubility of carbon dioxide

---

*The pH is an index which reflects the concentration of hydrogen ions; as the pH value decreases, the concentration of hydrogen ions and hence the acidity increases. The number of bicarbonate ions increases as a result of Eq. (2), but the relative change is small since the total concentration of bicarbonate ions is 2 mol/m^3 (Table 7.1). On the other hand, hydrogen ions have a considerable effect on the production of carbonic acid, firstly because at a pH of 8 the concentration of hydrogen ions (H^+) is a low 1×10^{-5} mol/m^3, and secondly because the concentration of hydroxyl ions (OH^-), which act to suppress the change in H^+ concentration, is also low (only 1×10^{-3} mol/m^3). Of course, the position of equilibrium in Eq. (3) shifts to the left in order to suppress the change in pH, but since carbonate ions are not present in sufficient quantities, the change in pH is not adequately suppressed.

increases; the cooler sea water can therefore dissolve considerable amounts of carbon dioxide from the atmosphere. The amount of carbon dioxide dissolved in sea water is measured, which allows the amount contained in the atmosphere at equilibrium to be calculated from the above equations; the actual concentration of atmospheric carbon dioxide is then subtracted, giving the ΔP value shown in Fig. 7.6 (in equatorial regions and in the eastern areas of the Pacific Ocean, this has a positive value). As the value of ΔP rises, it is expected that the amount of carbon dioxide released into the atmosphere from the oceans also increases.

However, the amount of carbon dioxide that is actually exchanged depends on the ease with which materials can be transferred at the interface between the surface sea water and the atmosphere; this is indicated by the value of the mass transfer coefficient.* After this value is also estimated, multiplication by ΔP permits an estimation of the actual amount transferred. An example of the results is shown in Fig. 7.7; in this case, the quantity transferred from the atmosphere to the ocean was calculated to be 1.6 gigatons. However, as was mentioned in Chapter 2, the actual amounts involved have not yet been clarified.

7.4 THE DISSOCIATION EQUILIBRIUM OF CALCIUM CARBONATE AND THE pH OF SEA WATER

Next in importance after carbonate ions are calcium ions, which are introduced into sea water following the weathering of terrestrial rock [Eqs. (2) and (4) in Chap. 5], and which are then removed due to the formation of calcium carbonate:

$$CaCO_3 = Ca^{2+} + CO_3^{2-} \tag{4}$$

$$Ca^{2+} + 2HCO_3^- = CaCO_3 + CO_2 + H_2O \tag{5}$$

Calcium carbonate exists in different forms, which have different properties. Although both calcite (a stable crystal) and aragonite (which has a semi-stable structure) are formed of calcium carbonate, aragonite has a greater solubility. Assuming that Eq. (4) (which represents the dissolution/formation of calcium carbonate) is in a state of equilibrium, then by also considering Eqs. (1)–(3) (but without considering alkalinity), calculations for a state of equilibrium with the current atmospheric concentration of carbon dioxide yield a pH value for sea water of 8.3, which is also approximately equal to the observed value. Under

---

*The amount of material transferred is directly proportional to the value of ΔP, with the constant of proportionality referred to as the mass transfer coefficient. The reciprocal of the mass transfer coefficient is termed the mass transfer resistance, and is the sum of the resistances of the ocean and the atmosphere. At the ocean/atmosphere interface, the mass transfer resistance of the atmosphere is dominant. However, when a strong wind blows, the mass transfer coefficient rises and the resistance falls.

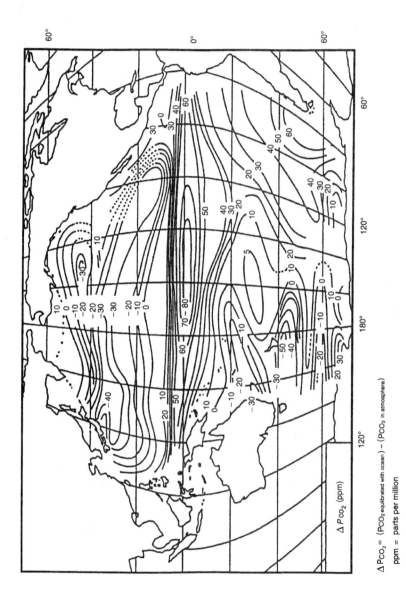

$\Delta P_{CO_2} = (P_{CO_2}$ equilibrated with ocean$) - (P_{CO_2}$ in atmosphere$)$

ppm = parts per million

FIGURE 7.6 CO_2 supersaturation in the waters of the Pacific ocean [Saruhashi and Sugimura, 1973, Miyake et al., 1974].

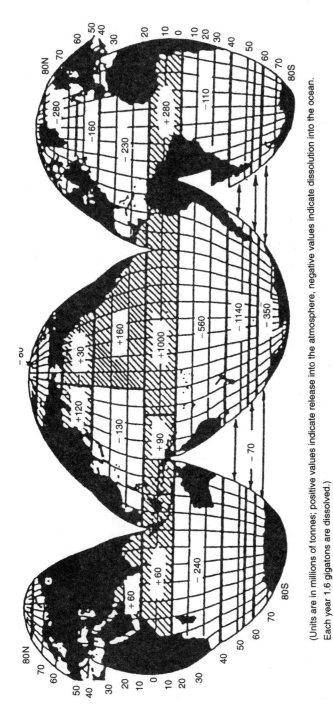

(Units are in millions of tonnes; positive values indicate release into the atmosphere, negative values indicate dissolution into the ocean. Each year 1.6 gigatons are dissolved.)

FIGURE 7.7   Estimated annual amount of carbon dioxide exchanged between the atmosphere and oceans [Takahashi, 1989].

these conditions, depending on Eq. (4), the ocean's capacity to absorb carbon dioxide is greater than that which would result by only considering Eqs. (1)–(3) at equilibrium. That is, when the concentration of atmospheric carbon dioxide rises and the amount of carbon dioxide dissolved in sea water increases by a large amount, hydrogen ions are produced [Eq. (2)], and the pH falls; however, in an attempt to suppress this change Eq. (3) then proceeds in the reverse direction (i.e. from right to left). But since the amount of CO_3^{2-} ions is originally small, the contribution of Eq. (3) is unlikely to be large unless CO_3^{2-} ions are produced by Eq. (4).

If calcium carbonate crystals are present, Eq. (4) proceeds from left to right with the production of carbonate ions, which in turn forces Eq. (3) from right to left. This causes a decrease in the value of the previously-mentioned buffer factor and an increase in the ocean's capacity to absorb carbon dioxide. (This is in fact exactly the same reaction as occurs in the weathering of limestone, which was discussed in Chapter 5.) Thus the calcium carbonate which is present in sea water aids the absorption of carbon dioxide. However, in reality this is the elution of carbon dioxide into sea water by calcium carbonate, and not the fixation of carbon dioxide as calcium carbonate.

When calcium carbonate is deposited (i.e. changes from a state of supersaturation to a state of equilibrium at the ocean surface), the process is the reverse of that described above, with the release of carbon dioxide from sea water. When deposition occurs, the number of carbonate ions decreases, which in turn leads to an increase in the concentration of hydrogen ions in accordance with Eq. (3). In order to compensate for this, Eq. (2) acts in reverse, causing Eq. (1) to also act in reverse, thereby releasing carbon dioxide. The end result is that carbon dioxide and calcium carbonate are formed from bicarbonate ions. A detailed investigation reveals that the extraction of 1 mole of calcium carbonate leads to the release of 0.6 moles of carbon dioxide, which is approximately the reaction described by Eq. (5).

The above argument at first appears to contradict the globally perceived view that excess carbon dioxide at the earth's surface is removed by the deposition and accumulation of limestone in sea water, but in actual fact it is possible to account for the apparent discrepancy by considering the weathering action. (For a more detailed explanation, please refer back to Section 5.4).

From what has been written so far, it might appear that the dissolution of calcium carbonate is in a state of equilibrium. However, in actual fact calcium ions are present in excess in the ocean's surface layers, which means that calcium carbonate can indeed be deposited; on the other hand, in the deep ocean the conditions are such that it dissolves. The main reason for this difference is the difference in pH. When the pH drops, the position of equilibrium in both Eqs. (2) and (3) shifts to the left, the number of carbonate ions decreases, and the calcium carbonate dissolves more readily. However, even if the pH drops suddenly, the reaction by which calcium carbonate compensates for this takes place extremely slowly, and thus the pH rises slowly. This is the situation which is progressively observed with increasing depth (Fig. 7.8). When the degree of saturation in Fig. 7.8

$$CaCO_3 \rightleftarrows Ca^{2+} + CO^{2-} \tag{4}$$

$$Ca^{2+} + 2HCO_3^- \rightarrow CaCO_3 + CO_2\uparrow + H_2O \tag{5}$$

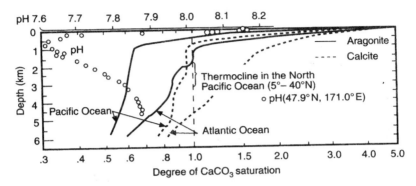

FIGURE 7.8 Changes with depth in pH and the position of equilibrium in the dissolution of calcium carbonate [Kojima, 1990, *adapted from* Tsunogai, 1985].

exceeds 1, deposition can occur; when it is less than 1, dissolution can take place. Since living organisms that make shells need to deposit calcium carbonate in order to make these shells, it is impossible for the organisms to live in the deep ocean when the degree of saturation is less than 1. It can thus be seen from Fig. 7.8 that calcite saturation occurs at a depth of around 1 km in the Pacific Ocean and at about 4.5 km in the Atlantic Ocean; for aragonite, saturation occurs at a depth of approximately 300 m in the Pacific and at about 2.5 km in the Atlantic.

The above discussion forms the basis for determining whether or not it is possible for the calcium carbonate to neutralize the acidity resulting from the dissolution of the liquefied carbon dioxide if the latter is disposed of in the deep ocean (as was discussed in Chapter 5). That is, even if liquefied carbon dioxide is injected into the oceans at depths greater than 3 km, it may be possible for the calcium carbonate (which *is* present at such depths in the Atlantic Ocean) to neutralize the effect of liquefied carbon dioxide. (In the Pacific, though, such hopes are faint.)

7.5 CHANGES IN CONCENTRATION OF PHOSPHORUS, NITROGEN AND CARBON WITH DEPTH, AND THE AGE OF THE WATER

The total amount of carbonic acid is greater in deep layers, and greater in the Pacific than in the Atlantic (Table 7.1). Let us now examine whether this is due to the fact that the waters of the Pacific are older.

$$P : N : C : Si = 1 : 16 : 106 : 16 \tag{6}$$

$$(CH_2O)_{106} \cdot (NH_3)_{16} \cdot (H_3PO_4) + 138\ O_2$$

$$\rightarrow 106 H_2CO_3 + 16 HNO_3 + H_3PO_4 + 16\ H_2O \tag{7}$$

$$P : N : C : Ca : Si = 1 : 15 : 120 : 40 : 50 \tag{8}$$

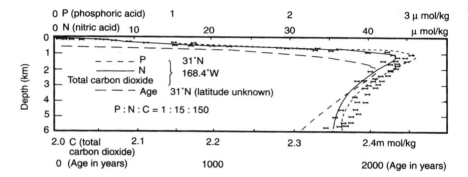

FIGURE 7.9 Pacific ocean: changes with depth in water age and concentrations of phosphoric acid, nitric acid and carbonic acid [Kojima, 1990, based on figures and data in Tsunogai, 1985].

Figure 7.9 shows the distribution with depth of both total carbon dioxide concentration and age, as well as the distributions of phosphoric acid and nitric acid. The left-hand edge of the horizontal axis represents a concentration of $2\ mol/m^3$, which is the concentration of carbonic acid in the ocean's surface layers; if the ratio between phosphorus, nitrogen and carbon ($P:N:C$) is taken as $1:15:150$,* then it appears that the actions of the three acids are virtually identical. Furthermore, there is close agreement with the pH trend shown in Fig. 7.8, suggesting that the origin of the change in pH might lie in the production of these acids.

7.6 ACTIONS OF LIVING ORGANISMS AND THE ELEMENTAL CYCLE

As mentioned earlier, specific chemical reactions occur as the age of sea water increases. But what actually causes them to occur?

$$P:N:C:Si = 1:16:106:16 \tag{6}$$

---

*It should be noted that although the figures given in this ratio [as well as in Eq. (8)] include Ca^{2+} and CO_3^{2-} (which form $CaCO_3$); this does not apply to Eqs. (6) and (7).

$$106H_2CO_3 + 16HNO_3 + H_3PO_4 + 16H_2O$$
$$= (CH_2O)_{106} \cdot (NH_3)_{16} \cdot (H_3PO_4) + 138O_2 \qquad (7)$$

$$P:N:C:Ca:Si = 1:15:120:40:50 \qquad (8)$$

The ratio of elements in phytoplankton (the so-called "Redfield ratio") is known to be that shown in Eq. (6). Accordingly, the photosynthesis reaction can be expressed by Eq. (7) (the reverse reaction which describes the process of decomposition and leads to the production of various acids). On the other hand, Eq. (8) shows the observed elemental ratio in particulate matter contained within sediments; the ratio is almost the same as that in Fig. 7.8. This composition of elements dissolves in water as an acid because of the decomposition of organic matter which either sinks or is transported as a result of the circulation of sea water.

On the other hand, as mentioned previously, the concentration of total dissolved carbon dioxide is roughly $2\,mol/m^3$ in the surface layers. In contrast, almost none of the phosphorus and nitrogen dissolve. Let us now consider the implications of this.

Apart from relatively rare localized areas which are artificially rich in nutrients, the oceans are actually poor in nutrients. (In fact, the sea tends to be nutrient-rich only in very limited areas such as estuaries and coral reefs).

In the oceans the density of living organisms is extremely low, and there is a lack of phosphorus and nitrogen (which are nutrients). On the other hand, carbonic acid is present in excess. Unfortunately, this means that even if the concentration of carbon dioxide in the atmosphere increases, and even if the concentration of total carbonic acid in the surface layers also increases, there will be no change in the number of living organisms in the oceans.*

A further point is that even if phosphorus and nitrogen are in plentiful supply, plankton are not formed in the surface layers because of the lack of trace elements such as iron and because of the extremely low temperature.

Plankton use the nutrient salts (i.e. salts such as phosphates and nitrates, which contain the nutrients of phosphorus or nitrogen) that rise gradually from the deep ocean through the thermocline. When the salts finally reach the depth at which sunlight can penetrate (a depth of approximately $100\,m$) the plankton photosynthesize and reproduce; they do not exist in very large quantities in the uppermost surface layers. Thus sunlight is not the limiting factor in the uppermost layers. In the regions where upwelling occurs, such as at the equator and in the coastal areas on the eastern sides of the oceans, nutrient-rich water rises from the deep ocean, brisk photosynthesis takes place, and the plankton reproduce, creating good fishing grounds. However, is carbon dioxide absorbed in such regions? If these waters, which are rich in phosphorus and nitrogen, were

---

*It should be pointed out that there is also an opposing view (for details, see Section 2.14).

pumped up and then brisk photosynthesis occurred, would this lead to the absorption of carbon dioxide?

The elemental composition of marine microorganisms is approximately constant in the ratio $P:N:C = 1:15:80-150$. Since phosphorus and nitrogen are formed as the result of the decomposition of living bodies after death, the carbon dioxide which is also formed as a by-product is present in excess (Fig. 7.9). When the deep layers upwell and the carbon dioxide reaches the surface layers, this carbon dioxide is released into the atmosphere when it reaches the surface layers. Therefore even if the organisms that use phosphorus and nitrogen engage in photosynthesis, absorption of carbon dioxide is unlikely. To put it another way, although the form of the elements is different at the bottom of the deep ocean, carbon is still present along with the phosphorus and nitrogen. If the above upwelling process proceeds a little too quickly, the excess carbon dioxide which is dissolved in the deep sea water will be released; however, there is no guarantee that the rate of photosynthesis due to the phosphorus and nitrogen will be sufficient for reabsorption of the released carbon dioxide. It is even possible that the net result will be the release of carbon dioxide, with photosynthesis lagging behind. In the final analysis, Fig. 7.8 unfortunately reveals that the raising of deep sea water and subsequent brisk photosynthesis is not an option for combating the problems associated with carbon dioxide.

Although the amounts concerned are minuscule, the phosphorus and nitrogen which are contained in materials such as fertilizers and detergents that flow into coastal regions could also be considered one of the missing sinks. However, all that can be said about the outflows of phosphorus from the natural world is that they form part of a vast process of geochemical circulation; that is, even in the ocean the phosphorus probably sinks to the sea bed as the remains of living organisms and later reappears at the earth's surface due to movements of the earth's crust.

Let us now consider nitrogen, one of the nutrients. The ocean fixes nitrogen from the atmosphere in the same manner as plants in the bean family. Along the Torres Straits of Papua New Guinea, the nitrogen-fixing capacity of blue-green algae is believed to be 3 kg of nitrogen per hectare per day. At tropical coral reefs blue-green algae covers the shallow ocean floor, with the annual rate of nitrogen fixation at such sites being as much as one tonne per hectare. This is an extremely large figure considering that the amount of nitrogen fertilizer applied to paddy fields each year is of the order of 100 kg/ha. However, on the other hand, in the ocean the majority of this is converted back to nitrogen, which is released into the atmosphere, with a small part being returned to the atmosphere as nitrogen oxide (which is another greenhouse gas).

A number of other elements are also present in various quantities, with varying recycling rates. In all cases, however, they pass through the cycle and end up in their original form. Thus, in many cases these elements could probably be considered to be in a steady state.

7.7 CARBON CYCLE MODELING IN THE OCEANS

The two-box model introduced in Chapter 2 is the simplest ocean-mixing model for calculating the rate of absorption of carbon dioxide by the ocean. It treats the surface layers and the deep ocean as two separate boxes in which thorough mixing occurs. By considering the decrease in the concentration of ^{14}C (an isotope of carbon) due to the carbon dioxide which is released from fossil fuels, and also the ratio of the two carbon isotopes ($^{12}C/^{14}C$) found in sea water, the model permits calculation of the rate of transfer between the atmosphere and the surface ocean layers, and also between the surface and deep ocean layers.

If it is postulated that the surface layers of the ocean are approximately 800 m deep, the theoretical and observed values for the amount of residual carbon dioxide in the atmosphere are in agreement. However, this assumption concerning the depth of the surface layers is contradicted by the fact that mixing in the surface layers is generally believed to only occur to a depth of about 100 m. That is, when the rate of absorption of atmospheric carbon dioxide by the ocean is estimated by an ocean-mixing model, the amount of absorption is found to be less than the amount of carbon dioxide which is removed from the atmosphere. The whereabouts of the missing carbon dioxide are unknown, and are referred to as "missing sinks".

One way of resolving this discrepancy is to revise the model. One such amended model is the "box diffusion" model*, which considers both a surface mixed layer and a deep diffusive layer. Other models have also included considerations of upwelling and downwelling flows. These modifications are attempts to incorporate into the model an accurate assessment of the speed with which exchange takes place between surface sea water and deep water. In order to assess the net transfer of carbon dioxide, besides the amount of carbon dioxide that is physically absorbed by the ocean, it is also necessary to consider the net fixation by living organisms and to incorporate this into future models. This entails a consideration of the net accumulation of calcium ions, phosphorus and nitrogen; their inflows from rivers etc.; and also their removal by sedimentation on the ocean floor. Naturally marine chemistry and temperature distributions must also be considered. It is also necessary both to use actual observations to clarify the ocean's behavior in detail, and to develop a highly precise quantitative model that incorporates all of the above-mentioned factors.

---

*This model recognizes that the deep layers of the ocean are not perfectly mixed, and that the concentration of carbon varies with depth; the transfer of carbon can then be calculated from the gradient of the concentration distribution.

7.8 PHYSICAL ABSORPTION OF ATMOSPHERIC
CARBON DIOXIDE BY THE OCEANS

Let us consider why the oceans physically absorb carbon dioxide, and the amounts that can be absorbed. The sea water that was in contact with the atmosphere before the industrial revolution, when the concentration of carbon dioxide was 280 parts per million (ppm) or less, accounts for a high proportion of the deep water today. That water is now slowly starting to appear at the surface of the ocean for the first time in one thousand years. During the intervening centuries, the concentration of atmospheric carbon dioxide has risen; the ocean therefore absorbs some of the carbon dioxide in an amount proportional to the increased atmospheric level. While almost negligible, there is also a contribution from other factors such as the increase in the inflow of phosphorus and nitrogen (due to the use of fertilizers and detergents etc.).

Although about sixty times the amount of carbon dioxide in the atmosphere is dissolved in the form of inorganic carbon, this does not mean that the carbon dioxide which is released from fossil fuels is distributed between the atmosphere and the ocean in the ratio of $1:60$. As was explained in Section 3, the buffer factor for the ocean has a value of approximately ten, resulting in a distribution ratio of only about $1:6$. Even so, this should mean that the proportion that remains in the atmosphere should be around one-seventh of the total.

In practice, however, this does not actually occur. The first reason that comes to mind is that the exchange between the atmosphere and the surface layers does not take place quickly enough. Nevertheless, taking the average between the values at the poles and at the equator, the situation appears to be approximately near equilibrium (although almost inevitably there is a divergence of opinion about this). If the surface layers were to be artificially agitated, it might be possible to approach equilibrium conditions and hence increase the absorption of carbon dioxide. While advantage could perhaps be taken of this as a means of combating carbon dioxide-related problems within certain localized areas, it is rather unrealistic, and its effectiveness is limited.

The main reason for the ineffectiveness is the poor transfer of materials between the surface and deep waters. For this reason, the upwelling of deep water still appears to hold some promise, even though an increase in photosynthesis due to phosphorus and nitrogen is unlikely (for reasons, see Section 7). Alternatively, if the global cycle were to be accelerated, and the mixing of deep and surface waters were to take place more quickly, the amount of physical absorption by the oceans would increase. Unfortunately, a simple calculation shows that a considerable amount of energy would be required to bring about transposition of the cold, heavy deep water and the warm, light surface layers; in addition, the amount of carbon dioxide released due to production of this energy would be far greater than the amount absorbed as a result of this countermeasure. This reason alone renders the technique basically unfeasible for fixing carbon dioxide.

Another proposal is to raise the deep water by placing barriers in the ocean currents in order to artificially create upwelling. This would also require approximately the same quantity of energy, but the difference would be that all of this would be supplied by the natural energy of the ocean currents. The water that would be brought up from the depths would be cold, whereas the water in the surface layers is warm. If this temperature difference could be used to generate electricity, then besides acquiring a source of energy, it should also be possible to achieve the secondary goal of absorbing carbon dioxide. This approach will be considered further in Section 12.

It is also necessary to consider the problem of photosynthesis, which is essentially the question of residence time of nutrient-rich water in the surface layers (that is, the length of time for which materials remain in these layers). If this time is not long enough for photosynthesis to proceed, the only result will be that the deep ocean, which is supersaturated with carbon dioxide, will become a source of carbon dioxide emissions. Consequently, this type of approach is fraught with danger in areas where downwelling occurs.

7.9 GROWTH OF MARINE ORGANISMS

Unfortunately, as mentioned previously, the net amount of photosynthesis would not increase even if the phosphorus- and nitrogen-rich deep water were brought up to the surface layers. This is because a corresponding amount of carbon dioxide would be formed due to the decomposition of dead organisms; the carbon dioxide would then dissolve in the sea water in excess, only to be released when the deep water reached the surface layers. To put it the other way round, the deep sea water contains an excess amount of dissolved carbon dioxide, which will be released; however, after this carbon dioxide is released, is expected that the same amount will eventually again be absorbed as a result of photosynthesis using the dissolved phosphorus and nitrogen.

The amount of biogenic carbon in the oceans is always controlled by the amounts of phosphorus and nitrogen, no matter whether the carbon is in the form of living or dead organisms, or whether these have decomposed and the carbon is present in the deep sea water in solution. That is, the amount of carbon present is always determined by the Redfield ratio. Based on this assumption, then irrespective of whether the water is in the deep or surface layers, an effective way to increase the amount of carbon originating from living organisms would be to add fertilizer to the oceans (the ocean fertilization technique). However, the fixation of one tonne of organic carbon would require about 0.2 tonnes of nitrogen and 0.02 tonnes of phosphorus; consequently, the fixation of all the carbon dioxide originating from fossil fuels would require over ten times the current world production of these materials.

At first glance this figure might seem substantial, but the amount of energy needed to synthesize fertilizer is actually not so large relative to the anticipated effects. (The energy expended in spraying the fertilizer would be negligible.) From this standpoint, therefore, it is definitely a feasible method for absorbing carbon dioxide. Calculations based on the $N:P:C$ ratio in marine organisms ($1:15:120$) show that during the manufacture of the nitrogen and phosphorus fertilizer needed to fix an annual amount of 5.0 gigatons of carbon, the amount of carbon dioxide that would be produced would be 0.33 gigatons of carbon equivalent. That is, the fixation of 100 units of carbon dioxide as organic marine matter would probably entail the production of an extra 7 units of carbon dioxide, which would be acceptable from the standpoint of energy resources. There would, however, appear to be a limit to the resources of phosphorus.

If it is accepted that the open seas are deficient in nutrients, then as long as fertilization is limited to those areas, the overall effect on the ocean environment is likely to be small. A simulation performed by the author and co-workers (Horiuchi *et al.*, 1994) showed that even if fertilization were carried out over a period of 100 years, the amount of organic matter contained in the surface layers would only increase by a factor of two. However, it is also necessary to investigate various environmental aspects and the optimization of spraying methods; this includes the time required for photosynthesis; diffusion and mixing between the surface and deep layers; the exchange processes that take place with the atmosphere. Attention would also need to be paid to the production of nitrous oxide (N_2O), the greenhouse gas which would be formed from the sprayed fertilizer.

The effectiveness of iron fertilization would also need to be evaluated, even though it poses few problems from the viewpoint of energy. This technique originally attracted attention because of the phenomenon observed in the Antarctic Ocean whereby the deep water first wells up to the surface and then sinks back down again. Because it is deep water, phosphorus and nitrogen are plentiful enough for photosynthesis to occur, but there is a shortage of iron since it easily precipitates and as a result is lost in sediments; it is therefore this element which limits the process. However, if iron were to be sprayed on to the oceans, it should lead to the same amount of carbon fixation as would have occurred with the anticipated degree of photosynthesis corresponding to the nitrogen and phosphorus contained in the sea water. Various problems remain, however, such as determining the best form of iron to employ. Experiments in the Pacific have demonstrated that a doubling of plant biomass can be achieved (Martin *et al.*, 1994).

Let us next consider the cultivation of seaweed etc. in the ocean itself. Assuming that the total weight of living organisms in the sea is determined by the quantity of phosphorus and nitrogen in the oceans, the amount of carbon present in the oceans on a global scale will not increase unless fertilizer is added to cultivate that seaweed (as explained in Chapter 5). If this seaweed is used as a source of soft energy, and the phosphorus and nitrogen within the living organisms again returned to the ocean, an artificial cycle will have been created, and a corresponding reduction in fossil fuel use will have been achieved. However, this technique needs to be evaluated not as a technique for the absorption of carbon dioxide but in terms of primary energy production, in the same manner as in the method by which plankton are cultivated in a tank using sunlight and with the use of highly concentrated carbon dioxide from sources of high emissions. (For details, please refer to Figs. 5.4 and 5.5).

The phosphorus and nitrogen which are contained in marine microorganisms along with water are present in much higher concentrations than in trees and shrubs etc. It has therefore also been proposed that these microorganisms should be used as fertilizers during afforestation. Once again, this suggestion requires both investigation as an energy system and also evaluation as a forestry system.

7.10   CORAL-GROWING TECHNIQUES

There is some debate as to whether coral plays a positive or negative role with respect to the problems associated with carbon dioxide. Over its entire history, in conjunction with weathering processes, the earth has either been neutral or else played a positive role with respect to carbon dioxide absorption. In contrast, from a purely chemical point of view, it has been concluded that coral acts as a source of carbon dioxide emissions [as expressed by Eq. (5) in this chapter] (for further details see Section 5.4). However, the debate has not yet arrived at a conclusion for a number of reasons.

Coral reefs are marine ecosystems which have a comparable (or even greater) capacity for primary production of organic matter than tropical rainforests (Table 6.1). That is, they absorb large quantities of carbon dioxide and use it to produce organic matter by means of photosynthesis. For this reason, coral reefs were at one time lauded for their contribution to solving the carbon dioxide problem.

There is, however, a counter-argument which holds that the productive capacity of the sea is controlled by phosphorus and nitrogen, and that if just these two elements are present, photosynthesis will proceed further; as a result, plankton will increase, and carbon dioxide should indeed be absorbed. However, it is further argued that if coral reefs are cultivated, the phosphorus and nitrogen which would normally go to other oceanic regions would instead be consumed by the coral reefs, and thus in the end no overall benefit would result.

In response, the advocates of coral reef cultivation argue that vegetation which is present in coral reefs, such as blue-green algae, brings about the fixation of atmospheric nitrogen, and that the coral reef ecosystem leads to the formation of organic matter using quantities of phosphorus and nitrogen which are much smaller than in the $P : N : C$ ratio that exists in other marine organisms.

The argument to refute this is the reverse of what was stated with respect to vegetation in Chapter 6, and is based on the fact that coral reefs retain only a small amount of organic carbon. The amount of carbon present in a coral reef ecosystem per unit area is one or two orders of magnitude smaller than in forests. The amount of organic carbon in coral reefs amounts to several hundred million tonnes, but this is only equal to about one month's emissions of carbon dioxide from fossil fuels. Therefore no matter how little phosphorus coral ecosystems use when organic matter is fixed, it soon undergoes decomposition and is released into the sea, and thus there is no overall effect when the oceans are considered on a global scale. Even so, it must still be recognized that a nitrogen-fixing effect does in fact contribute to the absorption of carbon dioxide.

However, if the organic matter which is produced by coral ecosystems and which contains only a small proportion of phosphorus and nitrogen does not undergo quick decomposition but instead sinks into the deep ocean, it could be argued that the effect would remain over a period of one thousand years (the time scale involved in one global cycle of the oceans). This point obviously demands clarification. From a purely inorganic chemical viewpoint, though, this is a negative result, and it is doubtful that the overall outcome is beneficial.

However, if coral ecosystems could actually produce a positive outcome, there is a chance that humans could eliminate the negative effects by artificially accelerating the weathering action [the opposite of the reaction described by Eq. (5)]. Further, even if coral reefs temporarily fail to yield positive results, the use of silicates as a source of calcium would mean that the only geochemical effect would be positive. This effect, and the question of which engineering techniques should be employed to attain these weathering effects, were topics which were addressed in Chapter 5.

Phytoplankton in the ocean engage in photosynthesis, which leads to the absorption of carbon dioxide; however, carbon dioxide is also released during the deposition of calcium carbonate. This release of carbon dioxide makes it necessary to revise upwards by about 30% the earlier estimate of the amounts of phosphorus and nitrogen required in the previously mentioned fertilization technique. Also, for the open ocean, Redfield et al. (1963) took phosphorus as the benchmark and concluded that the amount of nitrogen available for photosynthesis is only 94% of that required. If correct, this implies that nitrogen fertilization should be carried out rather than using phosphorus. On the other hand, the cultivation of coral in conjunction with the cultivation of forms of seaweed with a nitrogen-fixing ability might result in the coral cultivation being more effective if phosphorus (not nitrogen) fertilization were carried out.

7.11 ABSORPTION OF CARBON DIOXIDE BY THE OCEAN: THE OCEAN DISPOSAL OF HUMAN GARBAGE

As previously mentioned, the Pacific Ocean contains about 10% more dissolved carbon dioxide than does the Atlantic. Since the volume of the Atlantic Ocean is just less than 25% of the total volume of all the oceans, then by just increasing the concentration in the deep waters of the Atlantic by 10% (to make it equal to that in the Pacific), it would become possible to absorb virtually all of the carbon dioxide contained in the atmosphere. However, if ocean disposal of carbon dioxide were to be carried out (as discussed in Chapter 5), it would be essential to ascertain beyond any doubt both the most effective way of achieving this and the extent of the environmental problems which would be produced. Figure 5.11 clearly shows that the various effects of the ocean disposal of carbon dioxide would require investigation, e.g. the effect of dissolution and dispersion on marine environments (especially changes in pH), and the possibility of the regeneration of carbon dioxide and its subsequent re-release into the atmosphere. The honest truth is that too little is known about how the oceans function.

When the mechanisms at work in marine environments have been further clarified, it may be possible to make use of the existing descending flows in order to achieve the same results but without the expenditure of energy. Limestone serves as a possible example. Limestone dissolves in water which is supersaturated with carbon dioxide to form a solution of calcium bicarbonate (in the case of silicate-bearing rock, a greater degree of absorption takes place). When this is introduced into shallow water, the excess dissolved carbon dioxide is released (as previously described) and returned to the atmosphere. However, if use is made of the descending flows etc. and disposal in the deep ocean is carried out using the materials in the same form, the long period of retention in the ocean means that a large stock of carbon may be formed.

7.12 MULTIPLE USE OF THE OCEANS

Finally, let us consider the merits of adopting a system that makes multiple use of the oceans. As was previously mentioned, the generation of electricity due to temperature differences in the ocean is a technique which is related to the raising of deep water (although there are inherent economic problems). If the deep ocean water were pumped up, electricity could be generated from the temperature difference between the deep water and the surface water ("ocean thermal energy conversion"). A preliminary evaluation by the author and colleagues (Tahara *et al.*, 1994) showed that as a result of this electricity generation the consumption of fossil fuels would decrease by an equivalent amount, and carbon dioxide emissions would decrease. Furthermore, an amount of carbon dioxide equivalent to 10% of this reduction in carbon dioxide emissions would be physically absorbed by the deep water. Here again, though, it is necessary to consider the rates of diffusion and mixing in the surface layers.

When considering the problem of global warming in addition to the problems associated with carbon dioxide, it seems likely that the emergence of deep water in the surface layers would have a cooling effect on the earth. In practice, when electricity is generated from temperature differences in the ocean, heat in the surface layers is transferred to the deep water. However, the danger is that rather than merely warming the entire volume of water in the world's oceans, the effect will be to expand the volume occupied by the water, thus causing a rise in sea levels.

It is speculated that two major oceanic mineral resources in the ocean in the future will be manganese nodules (at depths greater than 4,000 m) and submarine hydrothermal mineral deposits (at a depth of 2,000 m). In addition there are high concentrations of copper, nickel and cobalt etc., which are present in over 100 times the amounts contained in underground deposits. Besides these, uranium is present in sea water at a concentration of 3 parts per billion (ppb) (about 4 gigatons are in solution, which is about 1,000 times the amount in underground reserves). The total amounts of dissolved gold, zinc, mercury, tin and silver are also estimated at between 100 and 1,000 times the amounts in land resources. Again, if these materials are extracted, a secondary effect will be that some of the deep ocean water will be raised to the surface. Since the deep water contains large amounts of phosphorus and nitrogen, this action will (as previously mentioned) not contribute to the solving of the problems related to carbon dioxide; however, in certain limited areas photosynthesis will proceed at a brisk rate, and it is possible that the artificial upflow will create good fishing grounds.

Methods for cultivating marine plant life are highly unlikely to lead to the absorption of carbon dioxide or to provide an alternative source of energy. However, the resulting biomass could be used as a fertilizer in agricultural areas, and if in addition it is used as fodder for livestock it will reduce the need for an equivalent amount of pasture land. If the pasture land is then replaced by regenerated forests, it will be an exemplary means of combating the adverse effects of carbon dioxide.

Besides using the ocean's resources and utilizing the natural energy contained in the oceans, it is hoped that an integrated approach involving a variety of systems will be adopted for the exploitation of the oceans. Of course, it should be remembered that the oceans do not belong to just oceanographers. The question of how human beings can use the oceans in an environmentally-friendly manner is a question that also requires input from individuals drawn from a diverse range of other backgrounds, such as agricultural specialists, engineers, economists and legal experts. In the final analysis, it is vital to have a multi-disciplinary evaluation from both the viewpoint of energy systems and also from the standpoint of environmental concerns.

References

Broecker, W. S. 1979. *Chemical Oceanography.* Harcourt Brace Jovanovich, Inc., USA. Citation from the Japanese version, p. 44, 1981, Tokyo Univ. Pub., Tokyo, Japan

Horiuchi, K. and T. Kojima, A. Inaba. 1995. *Energy Conversion Management* **36**, 915

Kojima, T. 1990. In Komiyama, H. *et al.* (eds.). 1990. *A Handbook of the Global Warming Issue*, p. 407, 426, 427. IPC, Tokyo, Japan (in Japanese)

Martin, J. H., K. H. Coale, K. S. Johnson *et al.*, 1994. *Nature* **371**, 123

Miyake Y., Y. Sugimura and K. Saruhashi. 1974. *Research in Oceanography Wks. Japan* **12**, 45

Redfield, A. C., B. H. Ketchum and F. A. Richards. 1963. The Influence of Organisms on the Composition of Sea-water. In Hill, M. N. (ed.). *The Sea: Ideas and Observations on Progress in the Study of the Seas.* Vol. 2 The Composition of Sea-water: Comparative and Descriptive Oceanography. pp. 26–77. Interscience Publishers, John & Sons, New York, USA

Saruhashi, K. and Y. Sugimura. 1973. *Chemistry* **28**, 770 (in Japanese). Cited in Kitano, Y., T. Matsuno. 1980. *Chemistry of the Earth and the Environment.* Iwanami Pub., Tokyo, Japan (in Japanese)

Stommel H. and A. B. Aron. 1960. *Deep-Sea Research* **6**, 140. Cited in Tomoda, Y., K. Takano. 1983. *Ocean*, p. 152. Kyoritsu Pub., Tokyo, Japan

Tahara, K., K. Horiuchi, T. Kojima and A. Inaba. 1995. *Energy Conversion Management* **36**, 857

Takahashi, T. 1989. *Oceanus* **32**, 22. Cited in Kaya, Y. (ed.). 1991. *A Handbook of Global Environmental Engineering*, p. 154, Ohm Pub. Inc., Tokyo, Japan (in Japanese)

Tomoda, Y. and K. Takano. 1983. *Ocean*, p. 151, Kyoritsu Pub., Tokyo, Japan (in Japanese)

Tsunogai, S. 1981. *Geochemistry* **15**, 70 (in Japanese). Cited in Tomoda, Y., K. Takano. 1983. *Ocean*, p. 156, Kyoritsu Pub., Tokyo, Japan

Tsunogai, S. 1985. *Chemical Key for Solve Oceanic Questions*, pp. 49, 62, 121, 132, Kyoritsu Pub., Tokyo, Japan (in Japanese)

Turekian, K.K. 1968. *Oceans.* Prentice Hall, Inc., Englewood Cliffs, New Jersey, USA. Citation from the Japanese version, 1971, p. 90, Kyoritsu Pub., Tokyo, Japan

8 THE CARBON DIOXIDE PROBLEM: A SUMMARY OF TECHNOLOGICAL MEASURES, AND A REVIEW OF POLICY AND ECONOMIC OPTIONS

8.1 SUMMARY OF TECHNOLOGICAL COUNTERMEASURES

A preliminary evaluation of the technological measures that can help combat the effects of increased carbon dioxide concentration is presented in Table 8.1. It is natural to proceed with alternative forms of energy, the development of energy systems, energy conservation and recycling etc.; indeed, there seem to be no other measures that can be implemented easily. Except for measures which involve theoretical problems or require overall evaluations as complete energy systems, the only courses of action which are feasible and on which research and development should proceed are the use of land vegetation; physical disposal and fixation in the oceans; the pumping up of deep water and use of ocean thermal energy; ocean fertilization; and utilization of geochemical recycling. Based upon this, let us consider what measures are possible from the policy and economic aspects.

8.2 THE DIFFICULTIES IN SETTING LONG-TERM NATIONAL TARGETS FOR REDUCING CARBON DIOXIDE EMISSIONS

As was mentioned in Chapter 1, the Intergovernmental Panel on Climate Change (IPCC) was established in November 1988, and published a report in September 1990. In parallel with this, in both the Netherlands (in November 1989) and at Geneva in Switzerland (November 1990) there was basically at least the appearance of agreement between the advanced nations regarding the stabilization of carbon dioxide emissions by the year 2000 (Fig. 8.1 summarizes each country's targets for curbing emissions). However, at the 'Earth Summit' in Rio de Janeiro in Brazil in June 1992, no progress at all was made in formulating the framework for a treaty, and indeed the negative attitude of the United States that was conspicuous. Subsequently the American presidential campaign unfolded, with Bill Clinton being elected president, and Al Gore (who was considered to be

Table 8.1 Summary of countermeasures against CO_2.

---

a) CO_2 countermeasures relating to the use of energy resources (Chaps. 3 and 4)
1. Energy conservation and substitution with alternative forms of energy.
2. Secondary energy systems must be developed in tandem with primary energy development. However, an overall evaluation is necessary, since it makes no sense to use CO itself as an energy medium.
3. Concentration of sources of CO emissions: It is not necessary to immediately proceed with technological developments on the basis of recovery and disposal.
4. Removal of carbon: The disposal of only carbon is unrealistic.
5. Use of low-carbon fuels: Considering the ability of the world's oceans to absorb CO_2 and the size of natural gas resources, it is possible that a rapid conversion now to low-carbon fuels could be counter-productive.

b) Recovery, utilization and storage of CO_2 (Chap. 5)
1. Recovery: Measures can be implemented using existing technology.
2. Conversion into resources, fixation, use: Conversion into resources requires more energy than is obtained when CO_2 is emitted, even if the CO_2 is used it will probably be re-released immediately. Consequently, this approach is without merit; care must be taken to avoid inadvertently pursuing this strategy. Furthermore, this overlaps with part a), section #2 above.
3. Ocean and subterranean storage: Technologically this would be possible, but it would be necessary to evaluate the effectiveness, safety and environmental impact.

c) Accelerated absorption of CO_2 by land vegetation (Chap. 6)
1. Fixation as lumber: Leads to a small reduction in CO_2.
2. Afforestation, and preventing the destruction of forests: While extremely important, this involves many serious social and economic problems.
3. Greenification of deserts: Deserts cover a vast area, and afforestation of these would also be extremely valuable in tackling the CO_2 problem; such greening of deserts would be technologically possible if only a sufficient amount of water could be supplied.

d) Accelerated absorption of CO_2 into the oceans (mainly Chap. 7)
1. Pumping up of deep water: due to the low concentrations of phosphorus and nitrogen in surface layers, surface layer photosynthesis would be activated by pumping up deep water. However, phosphorus and nitrogen are produced in the deep ocean by the decomposition of living organisms, and the CO_2 which is formed at this time is released at the surface, thus rendering the technique meaningless. Deep water is old and has a higher CO_2 absorption capacity; however, a substantial amount of energy is required for pumping up the water. It is therefore necessary to utilize ocean thermal energy (i.e. the energy due to temperature differences), etc.
2. Fertilization: If phosphorus and nitrogen are introduced to the nutrient-poor surface layers, the living organisms will increase in this area. In areas such as the Antarctic where the water upwells, inorganic phosphorus and nitrogen are abundant, but there is a shortage of iron; iron fertilization would be effective only in such exceptional cases, and it is necessary to evaluate the scale on which the technique could be applied.

Table 8.1 (*continued*)

3. Propagation of marine algae: The quantity of living organisms is determined by the amount of phosphorus and nitrogen, but since the same amount of decomposition occurs, the overall balance does not change. It would be effective to add fertilizer (including cultivation in tanks) and use this as a source of renewable energy and then recover the phosphorus and nitrogen; however, if the cost is high, fertilization only would be a more direct procedure. Implementation of this procedure in coastal areas would have a considerable environmental effect, and thus in a sense fertilization is the basic countermeasure.

e) Utilization of geochemical recycling (Chaps. 5 and 7)
1. Growing coral: This has value in that nitrogen is produced, but the accompanying chemical reactions would lead to the oceans becoming a source of CO_2:

$$Ca^{2+} + 2HCO_3^- = CaCO_3 + CO_2 + H_2O \tag{1}$$

2. Weathering: $$2CO_2 + CaSiO_3 + H_2O = Ca^{2+} + 2HCO_3^- + SiO_2 \tag{2}$$

Equation 2 alone suggest that 2 moles of carbon dioxide are absorbed by 1 mole of calcium silicate. But by combining the above two equations, it can be seen that the process leads to the overall recovery of 1 mole of CO_2. The reverse reaction of Eq. 1 (the weathering of limestone) also by itself leads to the isolation of CO_2. If this weathering action is induced and the products of the weathering action are disposed of in the deep ocean, no problem arises with ocean acidification, although a problem remains with the rate of the entire process.

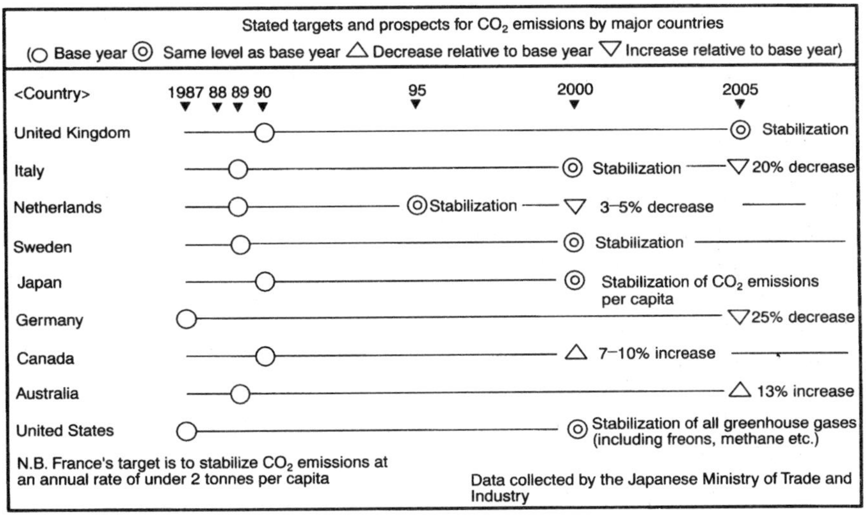

Stated targets and prospects for CO_2 emissions by major countries
(○ Base year ◎ Same level as base year △ Decrease relative to base year ▽ Increase relative to base year)

\<Country\>	1987 88 89 90	95	2000	2005
United Kingdom	○			◎ Stabilization
Italy	○		◎ Stabilization	▽ 20% decrease
Netherlands	○	◎ Stabilization	▽ 3–5% decrease	
Sweden	○		◎ Stabilization	
Japan	○		◎ Stabilization of CO_2 emissions per capita	
Germany	○			▽ 25% decrease
Canada	○		△ 7–10% increase	
Australia	○			△ 13% increase
United States	○		◎ Stabilization of all greenhouse gases (including freons, methane etc.)	

N.B. France's target is to stabilize CO_2 emissions at an annual rate of under 2 tonnes per capita

Data collected by the Japanese Ministry of Trade and Industry

FIGURE 8.1 National targets for CO_2 emissions [Kaya, 1991b].

pro-environment) becoming vice-president. This initially appeared to suggest that America would, like Western Europe, vigorously pursue reductions in carbon dioxide; however, these expectations have so far remained unfulfilled.

Although many countries claim to be seeking the stabilization or reduction of carbon dioxide emissions, this author feels that the commitments may be difficult to fulfill. Let us take the example of Germany, which is perhaps typical of Europe. This plan was formulated in 1987, so how was it possible to pledge as much as a 25% reduction in carbon dioxide emissions by the year 2005? An examination of the contents of Germany's own plans for meeting the target leads to the following points. Various factors must be taken into consideration when devising a plan, such as those presented in Fig. 3.3. First, what might be called prerequisites were certain levels of economic growth (measured by the gross per capita national product, which was expected to increase by 54%), and the population (which was expected to decrease by 1.3% by 2005). Under these conditions, in order to achieve a reduction in carbon dioxide it would be necessary to either reduce energy consumption with respect to GNP (a reduction in energy intensity), or else reduce carbon dioxide emissions as a percentage of energy consumption (a reduction in carbon dioxide intensity). As shown in Chapter 3, in order to reduce the emissions of carbon dioxide per unit consumption of energy it is necessary to convert either to fuel with a low-carbon content or else to nuclear power. However, conversion to nuclear power could well be difficult in the future, and is therefore perhaps unlikely. Again, a consideration of the amount of reserves shows that in the long term it is not possible to place great expectations in conversion to low-carbon fuels. Thus an estimated reduction in carbon dioxide intensity of 4.5% by 2005 (relative to the baseline year of 1987) was included in the plan. Of course if some epoch-making technology were developed for the recovery, disposal or fixation of carbon dioxide, this figure might increase further, but events so far do not make such a proposition seem likely.

A decrease in the energy consumption as a function of GNP requires either energy conservation or a change in industrial structure. The German target is a reduction of 47.9%, which is an annual rate of decrease of 3.6%. The conversion to industries with a low energy consumption essentially means transferring energy-intensive industries to neighboring countries; while this is inevitable, it is essentially meaningless from the point of view of the earth's environment. Ultimately, the burden must be relieved by energy conservation and greater efficiency. However, even in Japan after the 'oil shock' (1973–86) energy consumption as a function of GNP only decreased by an average annual rate of 2.8%; in addition, that figure includes changes in industrial structure. This demonstrates the immensity of the task involved in a 25% reduction, and raises doubts about Germany's ability to meet its stated objective.

Let us now consider future emissions in Japan. Table 8.2 shows a comparison between the forecasts by two Japanese bodies, The Institute of Energy

Economics and the Advisory Committee for Energy. Leaving aside the matter of which forecast is correct, there is a difference in the increase predicted for the consumption of non-fossil energy such as nuclear power. In comparison with the 1988 figure for carbon dioxide emissions (2.9 gigatons of carbon equivalent), the Institute of Energy Economics forecasts an increase to approximately 3.8 gigatons by the year 2000 (an increase of 1.30 times) and approximately 410 million tonnes by the year 2010 (an increase of 1.41 times). In contrast, the Advisory Committee for Energy predicts relatively smaller increases in emissions (1.16 times by the year 2000, and 1.18 times by 2010).

Unfortunately, there are doubts as to whether the rate of increase in alternative energy sources anticipated by the Advisory Committee for Energy is attainable, and thus the estimate by the Institute of Energy Economics appears more realistic.

According to Fujime (1991), in order to stabilize emissions in the year 2000 at current levels, it is necessary for the real economic growth rate of 4% predicted by the Advisory Committee for Energy to fall to 2%. A delay in implementing the stabilization period would lessen the effect of this decrease in growth rate on the economy. Also, if the stabilization occurred after the year 2000, it might be possible to avoid sacrifices by taking advantage of nuclear power and energy conservation.

However, even if the advanced nations (including Japan) were eventually made to stabilize emissions at current levels, this would only have a slight beneficial effect on the global environment. In the case of developing countries, predictions such as that shown in Fig. 8.2 show at a glance that an exponential growth in emissions may occur, which is a truly frightening proposition.

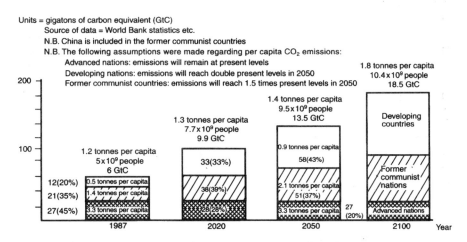

FIGURE 8.2 Predicted future CO_2 emissions from advanced nations, developing countries and former communist nations [Tomita, 1991].

The many difficulties involved in gaining agreement on realistic targets led to the failure of the 1992 Earth Summit to determine a strict framework for dealing with the problem; instead, participants merely followed a "pledge and review" approach.*

Table 8.2   Predicted Japanese energy supplies [Fujime, 1991].

Primary energy supplies	Actual figures	Institute of Energy Economics		Advisory Committee for Energy	
	1988	2000	2010	2000	2010
Coal (10^6 t)	115	132	132	142	142
Oil (10^6 kl)	263	348	383	308	306
Natural gas (10^6 t)	33	50	62	46	57
Hydroelectric power (10^9 kWh)	92	101	110	91	105
Nuclear power (10^9 kWh)	179	302	381	330	474
Geothermal energy (10^6 kl)	0.4	1.2	1.6	1.8	6.0
Alternative energy etc. (10^6 kl)	6.2	7.2	10.4	17.4	34.6
Total (10^6 kl)	467	621	697	597	666
Relative importance of energy sources					
Coal (%)	18.7	16.1	14.3	17.4	15.5
Oil (%)	56.4	56.0	54.9	51.6	46.0
Natural gas (%)	9.9	11.2	12.5	10.9	12.0
Hydroelectric power (%)	4.8	4.0	3.8	3.7	3.7
Nuclear power (%)	9.3	11.8	13.3	13.2	16.7
Geothermal energy (%)	0.1	0.2	0.2	0.3	0.9
Alternative energy etc. (%)	1.3	1.2	1.5	2.9	5.2

N.B. After excluding coke exports, totals are equal to 100%

*According to the so-called self-assessment approach, the policy is determined, and the results are self-evaluated. Consequently, there is no particular need to establish concrete objectives (including the balance between various countries), and naturally there would be no penalties.

Table 8.2 (*continued*)

Annual growth rate (%)	Institute of Energy Economics			Advisory Committee for Energy		
Year / Form of energy	1988– 2000	2000– 2010	1988– 2010	1988– 2000	2000– 2010	1988– 2010
Coal	1.1	0.0	0.6	1.8	0.0	1.0
Oil	2.3	1.0	1.7	0.9	0.0	0.5
Natural gas	3.5	2.2	2.9	2.9	2.1	2.5
Hydroelectric power	0.8	0.8	0.8	0.5	1.4	0.9
Nuclear power	4.5	2.3	3.5	5.2	3.7	4.5
Geothermal energy	10.0	3.2	6.9	13.3	12.8	13.1
Alternative energy etc.	1.3	3.7	2.4	9.0	7.1	8.1
Total	2.4	1.2	1.8	1.8	1.1	1.5
Economic growth rate (%)	3.70	2.90	3.34	4.00	3.00	3.54
E/G elasticity* ($-$)	0.65	0.40	0.55	0.45	0.37	0.42

*Ratio of growth rate in energy consumption to economic growth rate

N.B. The Advisory Committee for Energy calculated on the basis of overall supply of primary energy (production + imports).

The Institute of Energy Economics calculated on the basis of domestic supply of primary energy (production + imports − exports ± change in stocks).

8.3 LEGISLATION AS A POLICY OPTION

It seems impossible to directly restrict carbon dioxide emissions using legislation. A consideration of the balance between the various industries shows that it would be unrealistic for measures to cover the whole of the industrial world or all people. Nevertheless, it would be possible to implement constructive policies by establishing specific regulations for each industry and by instituting various other concrete measures.

Examples of these other measures would be restricting the driving of cars into cities, or limiting the days that cars could be used; tightening controls on illegal parking; instituting collection of refuse by type (together with penalties for non-cooperation) as a means for promoting recycling; introducing mandatory regulations for adiabatic structures to be employed in new buildings; and so on. However, if these measures were put in a package and then put to a referendum, they would probably be rejected. Also, if refuse were collected according to type for recycling, and at the same time a system of penalties instituted, would this inevitably mean that ordinary garbage would have to be put in transparent bags with the name of the person attached? Definitely not—although Tokyo planned to establish such a system in the mid-1990s; if this materialized, it would represent a loss of privacy and individual rights. (*Translator's note*: Following a public outcry over the proposal a scheme was instituted in 1995 whereby residents would use semi-transparent bags for combustible garbage without attaching their names.)

Whereas the above discussion concerned the imposition of regulatory measures, let us now consider the easing of restrictions. There appear to be several regulations which should be relaxed, such as the Electricity Utility Industry Law. Until very recently, if someone installed photovoltaic cells at their home, they were unable to sell the electricity and let another person use it, even if they had excess electricity. Similarly, it may be necessary to amend regulations in order to promote cogeneration.

However, it is necessary to carefully consider whether indirect regulation or the relaxation of laws and regulations can actually be of use in combating global warming and the problems posed by carbon dioxide. Any changes need to be thought through carefully before they are introduced.

As was mentioned earlier, any decrease in carbon dioxide would be meaningless if it were replaced by an increase in the levels of methane or nitrous oxide. Similarly, even if there was a drop in the atmospheric temperature, a problem would still remain if the sea level were to rise. Naturally the problem is not just one of laws and regulations; and the same applies to the economic measures that are discussed in the next section.

8.4 CARBON TAXES, ENVIRONMENTAL TAXES, AND ENERGY TAXES

A carbon tax is currently receiving attention as the simplest and most effective measure for combating the problem, and this tax would be levied according to the quantity of carbon dioxide emissions. The introduction of such a tax is certainly not just hypothetical; several countries have actually implemented one, including those in northern Europe (for the special features of these taxes, see Table 8.3). However, since it is rather difficult to fully comprehend the figures from the values per tonne of carbon, let us make a comparison with crude oil.

In 1992, the spot price of crude oil was $15 to $21 per barrel (1 barrel = 42 U.S. gallons = approx. 160 liters); this would be equivalent to around 10 cents per liter. If converted into the cost per kilogram of carbon it is a little more expensive, but let us assume the cost is 10 cents, which could possibly translate into a cost of about $100 per tonne. For gasoline in Japan, it would become $1.00 per liter, which would be of the order of $1,000 per tonne of carbon. Crude oil is extremely cheap, and even if a carbon tax were of the same order as the price of crude oil, at present price levels it would lead to gasoline prices going up by an extra 10 cents per liter at the outside.

Even now almost all countries levy taxes on energy. As will be mentioned later, the imposition of a tax of approximately $100 per tonne of carbon, which would probably have an economic impact, would be approximately the same as previous energy taxes, and would amount to only a little less than half of conventional oil taxes. For gasoline, the tax would translate into a price increase of at most 10%. The real value of a tax is that it forces industry to become energy-conscious. If this results in benefits for the environment, it should be implemented; however, the vital things to consider is whether a particular tax encourages the most desirable outcome.

Both Norway and Sweden have implemented income tax cuts, with no overall change in the balance of tax revenues (Table 8.3). Consequently, it is definitely possible to anticipate a slight effect on energy conservation. However, in the case of Sweden, the energy tax has been replaced by a carbon tax, under which the tax on high-carbon fuels has been increased, whereas it has been lowered on low-carbon fuels such as natural gas. But this serves no useful purpose. In the final analysis, the tax merely aims at promoting the use of low-carbon fuels.

Greater problems concern the tax exemptions for both peat* and biomass (even though most people might regard the latter exemption as natural). Such exemptions would of course act to promote conversion to these fuels. However, does this mean that carbon dioxide would not be generated by the use of peat? Is peat a renewable form of energy? Of course, the answer is 'no'. In addition, domestic biomass in Sweden is expensive, and biomass is therefore purchased from various foreign countries; and in some places it is used as a substitute for coal. Thus the argument for tax exemptions on these fuels seems dubious.

Let us examine another economic model. Figure 8.3 shows calculations for evaluating how effectively emissions could be abated by the imposition of taxes on the carbon equivalents, and includes a consideration of all greenhouse gas emissions (not only carbon dioxide). Since freon emissions are restricted by law, a reduction can be observed in total greenhouse gas emissions (which include freons); on the other hand, however, in order to stabilize carbon dioxide emissions, it would be necessary to impose a carbon tax of between $100 and $250 per tonne of carbon. Even for the same environmental tax, the effects depend on the time scale under consideration. Since we are considering a period of at least 100 years, the adopted policies must aim at greater efficiency, reduced energy consumption, the development of energy conservation measures, and the conservation and nurturing of plant life. It is necessary to have tax laws which promote

---

*Peat is formed when biomass accumulates in the earth and is the first step in its conversion into coal.

Table 8.3 Characteristics of carbon taxes introduced in Scandinavia and the Netherlands [Ogawa, 1992].

Country*	Sweden	Norway	Finland	Netherlands
Date of Introduction	January 1991	January 1991	January 1990	February 1990
Purpose of tax	—To strengthen indirect taxation by reforming the tax system —To promote energy conservation by raising prices	—To strengthen indirect taxation by reforming the tax system —To promote energy conservation by raising prices	—To strengthen indirect taxation by reforming the tax system —Part of an environmental damage tax	—To obtain funds necessary to implement environmental measures —Part of a fuel tax (environmental tax)
Tax rates	$152/tC	$128/tC(gasoline) $58/tC(mineral oil) $158/tC (natural gas)	$6.4/tC	$2/tC
% of total tax revenue	2×10^9 SKr Approx. 2.5%	2–2.5×10^9 NKr Approx. 1%	0.3×10^9 FIM Approx. 0.2%	0.15×10^9 Dfl Approx. 0.1%
Energy sources taxed	—All fossil fuels —Peat and methanol exempted	—Gasoline, mineral oils (oil for heating, light oils, heavy oils) —Offshore natural gas	—Fossil fuels except for those for transportation	—All fossil fuels

Expenditure of tax revenue	—General source of revenue —No effect on overall revenue (reductions in energy and income taxes)	—General source of revenue —No effect on overall revenue (reduction in income tax) —Appropriated for vehicular transport measures	—General source of revenue	—Appropriated for environmental measures
Tax exemptions and reductions	—Exemption for electrical power industry —To reduce energy-intensive industries	—Exemption for bunker oil used for coastal vessels and fishing boats		

*All countries provide tax exemptions for fossil fuels used as raw materials for the manufacture of chemicals, goods for export, bunker oil and aircraft fuel (the latter two are used for international transportation).

Skr = Swedish krone; NKr = Norwegian krone; FIM = Finnish marks; Dfl = Dutch guilders.

tC = tonnes of carbon.

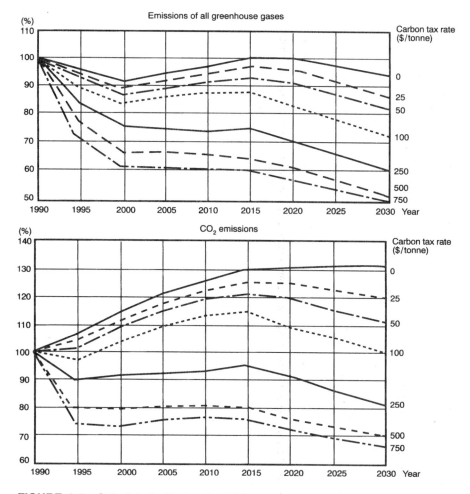

FIGURE 8.3 Calculated effects of a U.S. carbon tax on emissions of greenhouse gases [Matsuo, 1992, adapted from Department of Energy (DOE), 1991].

these goals; it is not acceptable to merely to convert from one fossil fuel to another.

The above discussion merely covered the effects of particular taxes. However, rather than just imposing taxes on consumption, it seems necessary to aim such taxes so as to promote investment in research into renewable energy (indeed, this is the most important area in which efforts must be expended); to fund the establishment of facilities for this purpose; or to facilitate policies that will encourage these outcomes. The world's environmental problems are becoming extremely serious; the only way to alleviate them is to develop sources of renewable energy over a period of at least several centuries. As stated previously, the problem surrounding carbon dioxide is essentially a problem of energy and resources.

Although the questions of lifestyle and recycling are also important, they cannot be dealt with only from the point of view of economics, and thus they will be discussed later.

8.5   AN "EMISSIONS MARKET" AND INTERNATIONAL CO-OPERATION

There are many views concerning the form in which an environmental tax should be introduced (Fig. 8.4).

In global terms, there would be no environmental improvement if one country reduces its carbon dioxide emissions only for another one to increase them. If a carbon tax came into effect only in the advanced nations, and if energy-intensive industries relocated to developing countries, then in the final analysis no change at all would have taken place on a global scale. Rather, if such industries relocated to countries with a lower technological standard, the overall effect would be to increase total carbon dioxide emissions. As a result, even if the situation is economically acceptable, problems remain with controlling carbon dioxide emissions on a global scale; the question then becomes one of how to deal with these additional problems.

The first idea is probably to impose an equal tax all over the world. However, it is likely that this would be rejected by developing countries—especially as developing countries use energy less efficiently. In addition, a tax would no doubt have an extremely harmful effect on their economies. According to one estimate, if a tax of several hundred dollars per tonne of carbon were introduced, the loss in terms of GNP would be many times greater, even though carbon dioxide emissions would be reduced. That is, in order to bring about a 1 tonne

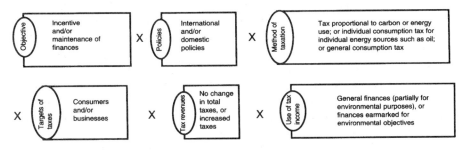

FIGURE 8.4   Issues involved in environmental taxes [Matsuo, 1992].

decrease in carbon, the GNP would suffer a loss of close to $3,000. In addition (and this is a greater problem) it is possible that carbon dioxide emissions might not actually fall, and the end result would simply be that the economy would be thrown into turmoil.

This argument led to the concept of establishing a market in tradeable emissions permits, whereby permits to emit given amounts of carbon dioxide could be bought and sold. The first problem that arises with this proposal is the way of determining how permits would be distributed around the world. For example, let us suppose that the permits are distributed everywhere on the basis that each person in the world has an equal right to emit carbon dioxide. Then an area which wishes to purchase permits levies a charge on the carbon dioxide that it itself produces, and uses that revenue to purchase a permit. Figure 8.5 shows the results of such a calculation for a total carbon dioxide emission of 5 gigatons of carbon equivalent in the year 2000. Since Africa and south-east Asia have an excess of permits, they do not need to impose any taxes but are able to receive revenue from the sale of permits. On the other hand, advanced nations such as the United States would impose charges of over $100 per tonne on carbon dioxide emissions, and would purchase permits at a cost of $215.4 per tonne of carbon. This means that the emitters of carbon dioxide would pay less than in the plan to curtail emissions by taxation only.

Figure 8.6 shows this scenario recalculated to allow for the effect of the fixation of carbon dioxide which would occur with afforestation. Of course, numerous problems would arise regarding matters such as whether afforestation will actually lead to the fixation of carbon dioxide, as well as details concerning administrative structures etc. Also, the cost of afforestation would vary from country to country and from region to region. In addition, there is a wide variation in opinion regarding estimates of the areas in which afforestation is feasible. Afforestation would probably first be carried out slowly in the areas where it is easiest to achieve; the present calculation assumed that the maximum rate of afforestation would be 1% of the available area per annum. Based on an initial cost of afforestation of $4 to $15 per tonne of fixed carbon, it has been calculated that the cost would increase to 10 times that when 25% of the afforestation had been carried out, and that when the process was completed the cost would, depending on the area under consideration, have risen to nearly $1,000. This would mean that the previously mentioned cost in the tradeable permits market of $215 by the year 2000 would drop to a mere $25. Of course, this calculation assumes that afforestation could be accomplished relatively cheaply.

A comparison of the results (Table 8.4) shows that by including afforestation the market can be realized in 2050 with each country able to achieve economic growth, even when no country is able to export emissions permits (i.e. when total emissions are restricted to 4 or 5 gigatons of carbon). That is, regulations can succeed in meeting the objective, provided that the whole process is managed well.

Table 8.4 Calculated costs of emissions permits in terms of 1975 US dollars per tonne of carbon [Okada, 1993].

Year	2000 A.D.			2050 A.D		
Total regulated CO_2 emissions (GtC)	4.0	5.0	6.0	4.0	5.0	6.0
Supply and demand for permits in balance	586.7	212.8	42.3	(a)	(a)	989.9
Introduction of an option for CO_2 absorption by afforestation	80.8	25.2	11.2	1190.5	418.7	213.7

(a) Regions selling emissions permits will disappear, and since it is impossible to establish an international emissions permit market, those regions that exceed their initial CO_2 emissions quotas will have to use taxation to reduce CO_2 emissions to the level of their initial quotas.
GtC = gigatons of carbon

Of course, it could be argued that all this is merely an unproven theory. For example, questions arise as to when international agreement would be forthcoming. However, when all things are considered, afforestation is the only feasible countermeasure that is possible at the present time. Nevertheless, it is surely an approach that needs to be investigated thoroughly.

8.6   A FRUGAL LIFESTYLE IN AN ERA OF FEW GOODS

Finally, let us consider a lifestyle in which little energy is used, and recycling is common. To protect the earth we need to take positive actions, but as stated there appear to be many difficulties if only economic measures are adopted. This would suggest that instead of just taxation and subsidies, it is also necessary to consider the problem from the ethical standpoint. Even today, we cannot just throw garbage anywhere; in the same way, the time might possibly come when we must first sort all garbage by type and not throw away anything that can be used again (which implies that we should not buy, receive or give items that will eventually become garbage).

We all use an excessive amount of energy, and this appears to represent a mistaken sense of values. Unfortunately, the saving of energy frequently requires the loss of time. We therefore need to slow down and take more time to think carefully about the consequences of our actions, and act in a manner that will allow coming generations to inherit an unspoiled environment which is blessed with fully conserved resources.

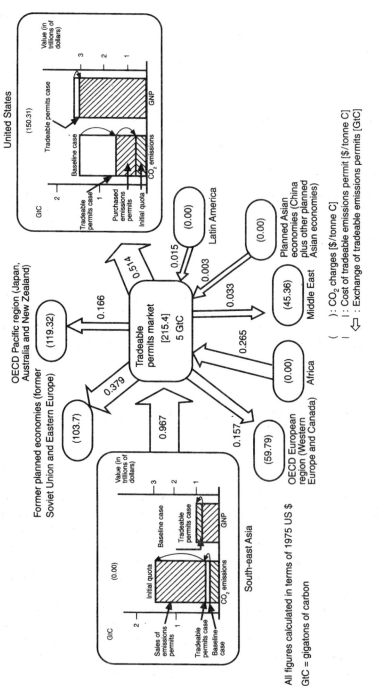

FIGURE 8.5 Market for tradeable emissions permits in the year 2000 assuming a total emission of 5 gigatons of carbon [Okada, 1991].

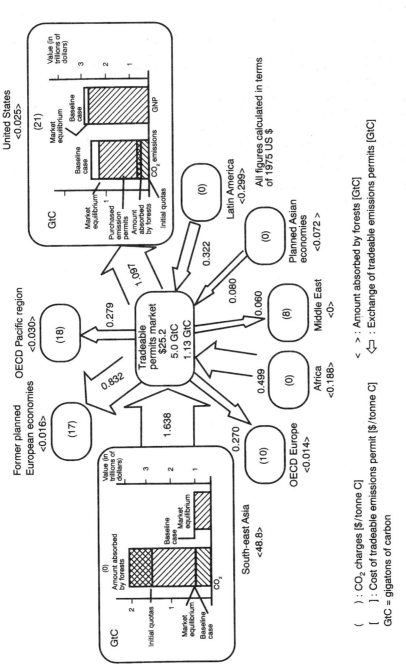

FIGURE 8.6 Market for tradeable emission permits in the year 2000 assuming a total emission of 5 gigatons of carbon with an allowance for CO_2 absorption by afforestation [Okada, 1991].

References

United States Department of Energy (DOE). 1991. *Limiting Net Greenhouse Gas Emissions in the United States—Report to the Congress of the United States.* U.S. Department of Energy F8. Cited in Fujime, K. 1991. Power **41** (205), 5 (in Japanese)

Kaya, Y. 1991. *System, Control and Information* **35**, 529 (in Japanese)

Matsuo, N. 1992. *Petrotech* **15**, 947 (in Japanese)

Ogawa, Y. 1992. *Petrotech* **15**, 817 (in Japanese)

Okada, K. and K. Yamaji. 1991. *Power Economics Research* **29**, 11

Okada, K., H. Yamamoto, K. Nagano, K. Yamaji and K. Nagata. 1993. *Energy and Resources* **14**, 48 (in Japanese)

Tomita, A. 1991. Coal Utilization in Global Environment—Coal and CO_2. p. 2. Report of Grant-in-Aid for Scientific Research, The Ministry of Education Science and Culture, Japan (in Japanese)

BIBLIOGRAPHY

For interested readers who wish to study the issues raised in this book in greater detail, the following is a list of journals, books and conferences that may provide a source of the latest research findings and current thought in the field. Recent international conferences are listed in chronological order; if the conference forms one of a series of regular conferences, then only the most recent is listed.

Related Journals

Chemosphere
Climatic Change
Energy
Energy and Fuels
Energy Conversion and Management
Energy Policy
Environmental and Resource Economics
Geophysical Research Letters
Global Biogeochemical Cycles
International Journal of Climatology
Journal of Air and Waste Management Association
Journal of Climate
Journal of Geophysical Research
Journal of Power and Energy
Nature
Oceans
Power
Science
Tellus
The Energy Journal
Water, Air and Soil Pollution

Carbon dioxide problems in general (including both emissions and abatement)

Intergovernmental Panel on Climate Change (IPCC) by WMO (World Meteorological Organization), UNEP (United Nations Environment Programme). 1990. J. T. Houghton, G. J. Jenkins, J. J. Ephraums (eds.). *Climate Change: The IPCC Scientific Assessment.* Cambridge University Press, Cambridge

Intergovernmental Panel on Climate Change (IPCC). 1992. J. T. Houghton, B. A. Callender and S. K. Varney (eds.). Climate Change 1992: The Supplementary

Report to the IPCC Scientific Assessment, Cambridge University Press, Cambridge

Intergovernmental Panel on Climate Change (IPCC). 1996. Climate Change 1995: The Science of Climate Change. J. T. Houghton, L. G. Meira Filho, B. A. Callender, N. Harris, A. Kattenberg and K. Maskell (eds.). Cambridge University Press, Cambridge

Smith, I. M., C. Milsson and D. M. B. Adams. 1994. *Greenhouse Gases— Perspectives on Coal*. IEAPER/12. IEA Coal Research, London

Smith, I. M., K. Thambimuthu, 1991. *Greenhouse Gases—Abatement and Control*. IEACR/39. IEA Coal Research, London

Global environmental issues (including climate change and the carbon cycle) (Chapters 1 and 2)

Brown, L. R. *et al.* 1995: *State of the World 1995*. (World Resources Institute). W. W. Norton & Company, Inc., New York

Erickson, J. 1990. *Greenhouse Earth—Tomorrow's Disaster Today*. TAB Books, Blue Ridge Summit, Pennsylvania

IUCN (The World Conservation Union), UNEP (United Nations Environment Programme), and WWF (World Wide Fund for Nature). 1991. *Caring for the Earth—A Strategy for Sustainable Living*. Gland, Switzerland

Meadows, D. H., D. L. Meadows and J. Randers. 1992. *Beyond the Limits— Confronting Global Collapse, Envisioning a Sustainable Future*. Chelsea Green Pub. Co., Vermont

Meadows, D. H., D. L. Meadows, J. Randers and W. W. Behrens III. 1972. *The Limits to Growth—A Report for The Club of Rome's Project on the Predicament of Mankind*. Universe Books, New York

World Commission on Environment and Development. 1987. *Our Common Future*. Oxford University Press, Oxford

Scientific Committee on Problems of the Environment (SCOPE). 1981. Bolin B. (ed.). *Carbon Cycle Modelling*. John Wiley & Sons, New York

The 7th Global Warming International Conference (GW7), Vienna, Austria, April 1–3, 1996

Energy and Economics (Chapters 3, 4 and 8)

Boyle, G. (ed.). 1996. *Renewable Energy—Power for a Sustainable Future*. Oxford University Press, Oxford, in association with the Open University, Milton Keynes

Scheer, H. 1994: A Solar Manifesto, James & James (Sci. Pub.), London

Vernon, J. 1992. Carbon Taxes, IEAPER/01, IEA Coal Research, London

International Symposium on Energy, Environment and Economics. Australia, November 20–24, 1995

Joint IEW (International Energy Workshop)/JSER(Japan Society of Energy and Resources). International Conference on Energy, Economy, and Environment. Osaka, Japan. June 25–27, 1996

31st Intersociety Energy Conversion Engineering Conference, Washington D.C., August 11–16, 1996

International Energy & the Environment Conference, Sydney, Australia, August 28–29, 1996

3rd International Conference on Renewable Energy. Asia Pacific '96. Exhibition & Conference. Manila, Philippines, October 7–9, 1996

4th International Conference on Technologies & Combustion for a Clean Environment, Lisbon, Portugal, July 7–10, 1997

Urban Transport & the Environment for the 21st Century. Barcelona, Spain, October 2–4, 1996

8th International Conference on Coal Science, Essen, Germany, September 7–12, 1997

Carbon dioxide recovery and disposal (Chapter 5)

Hendricks, C. 1994. *Carbon Dioxide Removal from Coal-Fired Power Plants.* Kluwer Academic, Dordrecht, Netherlands

1st International Conference on Carbon Dioxide Removal. Amsterdam, Netherlands, March 4–6, 1992. Energy Conversion Management 33 (5–8), 1992

International Energy Agency (IEA). Carbon Dioxide Disposal Symposium, Oxford, March 29–31, 1993. Energy Conversion Management 34 (9–11), 1993

International Energy Agency (IEA). Greenhouse Gases: Mitigation Options Conference, London, August 1995. Energy Conversion Management 37 (6- –8), 1996

2nd International Conference on Carbon Dioxide Removal, Kyoto, Japan, October 24–27, 1994. Energy Conversion Management 36 (6–9), 1995

3rd International Conference on Carbon Dioxide Removal, Massachusetts, USA, September 9–11, 1996. To be published in Energy Conversion Management

International Workshop on Greenhouse Gas Mitigation Technologies & Measures, Beijing, China, September 9–12, 1996

International Symposium on Ocean Disposal of Carbon Dioxide, Ocean Disposal, Tokyo, Japan, October 31–November 1, 1996

4th International Conference on Carbon Dioxide Utilization (ICCDU4), Kyoto, Japan, September 7–11, 1997

The Function and Use of Plants and Soils (Chapter 6)

Hillel, D. 1971. *Soil and Water.* Academic Press, Inc., San Diego, USA

Kuroda, Y. and F. Nectoux. 1989. *Timber from the South East*. World Wide Fund for Nature (WWF). International, Switzerland

Mather, A. S. 1990. *Global Forest Resources*. Pinter Pub., London

Westoby, J. 1989. *Introduction to World Forestry—People and their Trees*. Basil Blackwell Ltd., Oxford, UK

Whittaker, R. H. 1975. *Communities and Ecosystems* (2nd Ed.). Macmillan, New York, USA

Wild, A. 1993. *Soils and the Environment*. Cambridge University Press, UK

Desert Technology IV (Engineering Foundation Conference), Kalgoorlie, Western Australia, September 22–26, 1997. To be published in Journal of Arid Land Studies

Role and Use of the Ocean (Chapter 7)

Broecker, W. S. 1979. *Chemical Oceanography*. Harcourt Brace Jovanovich, Inc., USA

Hill, M. N. 1963. *The Sea—Ideas and Observations on Progress in the Study of the Sea*. John Wiley & Sons, New York

Turekian, K. K. 1968. *Oceans*. Prentice Hall, Inc., Englewood Cliffs, New Jersey, USA

Von Arx, W. S. 1962. *An Introduction to Physical Oceanography Reading*. Addison Wesley, Massachusetts, USA

INDEX

221